RESETTING THE KITCHEN TABLE: FOOD SECURITY, CULTURE, HEALTH AND RESILIENCE IN COASTAL COMMUNITIES

RESETTING THE KITCHEN TABLE: FOOD SECURITY, CULTURE, HEALTH AND RESILIENCE IN COASTAL COMMUNITIES

CHRISTOPHER C. PARRISH
NANCY J. TURNER
SHIRLEY M. SOLBERG
EDITORS

Nova Science Publishers, Inc.
New York

For permission to use material from this book please contact us:
Telephone 631-231-7269; Fax 631-231-8175
Web Site: http://www.novapublishers.com

NOTICE TO THE READER

The Publisher has taken reasonable care in the preparation of this book, but makes no expressed or implied warranty of any kind and assumes no responsibility for any errors or omissions. No liability is assumed for incidental or consequential damages in connection with or arising out of information contained in this book. The Publisher shall not be liable for any special, consequential, or exemplary damages resulting, in whole or in part, from the readers' use of, or reliance upon, this material.

Independent verification should be sought for any data, advice or recommendations contained in this book. In addition, no responsibility is assumed by the publisher for any injury and/or damage to persons or property arising from any methods, products, instructions, ideas or otherwise contained in this publication.

This publication is designed to provide accurate and authoritative information with regard to the subject matter cover herein. It is sold with the clear understanding that the Publisher is not engaged in rendering legal or any other professional services. If legal, medical or any other expert assistance is required, the services of a competent person should be sought. FROM A DECLARATION OF PARTICIPANTS JOINTLY ADOPTED BY A COMMITTEE OF THE AMERICAN BAR ASSOCIATION AND A COMMITTEE OF PUBLISHERS.

Library of Congress Cataloging-in-Publication Data
Available upon request.

ISBN 10 1-60021-236-0
ISBN 13 978-1-60021-236-9

Published by Nova Science Publishers, Inc. ✦New York

CONTENTS

FOREWORD

The results of the work of *Coasts Under Stress* (*CUS*) are to be found in numerous journal articles, two films, one booklet, one book, and four edited collections, showing how the various parts of life in coastal communities fit together and how interactive restructuring has generated the risks, threats, and opportunities coastal communities (human and biophysical) confront. Three of the team books are theme-based. One is on social-ecological knowledge systems and on the vital importance and challenges associated with moving knowledge across disciplinary boundaries, within, and between knowledge systems and from people to researchers to policy-makers to students and back to communities in order to grapple with interactive restructuring and its effects. One is on the relationship between interactive restructuring and power, whether as energy (oil and gas, hydro) or as "power over nature constructs," or as power and agency in nature and human communities. This one is on the history of health, diet, and nutrition—with a particular focus on the issue of decreasing food security in places where once-stable food webs have suffered radical shock, as have the cultures of human communities, which have always been interdependent with now-endangered food sources. There are two publications for special audiences: one for policy-makers and one for coastal communities. The principal volume was team-written and is, in itself, an experiment in interdisciplinary scholarship. The team planned the volume, contributed their findings to all the sections, commented on the manuscript as it evolved, and have approved its final shape.

In all our work, by *environmental restructuring* we mean changes in the environment, usually at large scales, which are thought to be caused, at least in part, by such things as climate change. We take *social restructuring* to mean changes in society at a range of scales that result in, for example, changes in community cohesion, social support, health care delivery, or the availability of educational resources. Such changes include industrial restructuring which deals with changes in patterns of ownership and control and in work environments, and political restructuring which deals with shifts in policy regimes. We take *health* to be the capacity to cope with stressors and recognizing that people are a part of (not outside) nature. *Social-ecological* health is the capacity of the human–natural world nexus to deal resiliently with change and the stress that brings.

We wish to take this opportunity to thank the Social Sciences and Humanities Research Council of Canada (SSHRC), The Natural Sciences and Engineering Research Council of Canada (NSERC), Memorial University of Newfoundland, and The University of Victoria for major funding of this work, and for ongoing support throughout the lifetime of the Project.

We owe a debt of gratitude to M. Yves Mougeot and Mme. Katharine Benzekri of SSHRC, along with the various SSHRC officers who assisted us, particularly Jacques Critchley who got us started, Pierre-Francois LeFol who was with us in our "middle period," and Michele Dupuis who has seen us through to the end. We also wish to express our gratitude to André Isabelle and Anne Alper of NSERC, whose assistance has likewise been invaluable throughout all the years of our work. This project could not have been carried out without a dedicated staff, and we here thank Janet Oliver, Carrie Holcapek, Cathy King, Kari Marks, Angela Drake, and Moira Wainwright for their hard work, constancy, and continued support through thick and thin. In addition, Carrie Holcapek worked tirelessly on getting this volume into shape. We wish also to thank the other universities whose faculty contributed to our work: The University of British Columbia (and, in particular, the Fisheries Centre and the Department of Geography), Dalhousie University, Saint Mary's University, and the University of New Brunswick. Our heartfelt thanks goes to our partners and our Advisory Boards, named in the Appendices to this volume, and to the Centre for Studies in Religion and Society and the Centre for Earth and Ocean Research, both at the University of Victoria, for providing the west coast part of the team with a home. On the East Coast, Memorial University provided a small building for the use of staff, faculty and students, while on the West Coast the University of Victoria gave the Project Director an academic home. We are grateful to both these institutions for their generosity and for their faith in us.

Rosemary E. Ommer
University of Victoria, January 2006

ABBREVIATIONS

BC   British Columbia, Canada
C   Celsius
COSEWIC   Committee on the Status of Endangered Wildlife in Canada
CPUE   Catch per unit effort
CUS   *Coasts Under Stress* research project
DFO   Department of Fisheries and Oceans, Government of Canada
DHA   Docosahexaenoic acid
DRI   Dietary reference intakes
EPA   Eicosapentaenoic acid
FAO   Food and Agriculture Organization of the United Nations
FFA   Free fatty acids
g   Gram
GMO   Genetically modified organism
HBC   Hudson's Bay Company
IFFO   International Fishmeal and Fish Oil Organisation
ITQ   Individual transferrable quota
kg   Kilogram
LEK   Local ecological knowledge
m   Metre
MDS   Multidimensional scaling
mg   Milligram
MPA   Marine protected area
NAFO   Northwest Atlantic Fisheries Organization
ng   Nanogram
NL   Newfoundland and Labrador, Canada
NSERC   Natural Sciences and Engineering Research Council of Canada
PCA   Principal components analysis
PCB   Polychlorinated biphenyl
ppb   Parts per billion
psu   Practical salinity units
PUFA   Polyunsaturated fatty acids
SCOPE   Scientific Committee on Problems of the Environment
SD   Standard deviation
SSHRC   Social Sciences and Humanities Research Council of Canada

TEK	Traditional Ecological Knowledge
UBC	The University of British Columbia
UN	United Nations
US FDA	United States Food and Drug Administration

In: Resetting the Kitchen Table
Editors: C. C. Parrish et al., pp. 1-12

ISBN 1-60021-236-0
© 2008 Nova Science Publishers, Inc.

Chapter 1

INTRODUCTION—FOOD SECURITY IN COASTAL COMMUNITIES IN THE CONTEXT OF RESTRUCTURING: COASTS UNDER STRESS

Nancy J. Turner[1], Christopher C. Parrish[2] and Shirley M. Solberg[3]
[1]Environmental Studies, University of Victoria, Canada
[2]Ocean Sciences Centre, Memorial University of Newfoundland, Canada
[3]Nursing, Memorial University of Newfoundland, Canada

ABSTRACT

Food security—access to sufficient, safe, nutritious, and culturally appropriate food—for all peoples of the world is a goal of the World Food Summit and other international bodies, but it is an elusive target, even in developed countries. Rural communities, including those on the Atlantic and Pacific coasts of Canada, are struggling to retain access to healthy foods. Many of the local food resources on which people relied in the past—for example, cod on the East Coast and salmon on the West Coast—are far less plentiful today than in the past, largely due to over-exploitation and environmental deterioration. In this chapter we discuss the complex linkages among environmental change, economic restructuring, ecosystem health, and human health relating to food production and availability in coastal communities. We first outline the basic nutritional requirements for humans and then use some key examples, from the antiscorbutic properties of spruce beer to the production and nutritional importance of omega-3 fatty acids, to demonstrate how diet and lifestyle affect health in these communities. Our book title, "Resetting the Kitchen Table," alludes to the widespread shifts in food production, processing, storage, and marketing that have occurred within Canadian coastal communities, and the effects of these changes on peoples' lifestyles, nutrition, and health. Concerns with environmental contaminants in food, maintaining access to healthy food at reasonable cost, and retaining the knowledge associated with local food production are all factors affecting food security in these communities. This chapter also provides a framework for the rest of the sections and chapters in the book.

INTRODUCTION

Food security exists when all people, at all times, have physical and economic access to sufficient, safe, nutritious and culturally appropriate food to meet their dietary needs and food preferences for an active and healthy life (World Food Summit 1996).

… Our children, our young people, because they're employed, they're not able to come down here and spend enough time here to observe to learn how to survive in this camp [Gitga'at seaweed camp at Kiel, north coast of British Columbia]. They have to be taught. The food resources . . . we're rich in food resources and they're acquiring a taste that is different than ours and so we have all these nutrients, all the iron, all everything that you could want in the food that we have here, between the meat, and the sea, and the shellfish you have everything that you could possibly want for your body, for your use, that you don't have to settle for hamburger and bologna, and Ichiban, and macaroni, and Kraft dinner. We have everything that we want. I've always said that we're rich; we're rich in our food (Helen Clifton, Gitga'at Nation, Hartley Bay, BC, 2002).

Access to good nutrition is a basic human need, and food security is recognized as a critical issue facing humans all over the globe (Hulse 1995). Food production and availability are dependent on many factors, both social and environmental. In Canada's coastal communities, as in many communities throughout the world, economic restructuring has often compromised the ability of people to meet their nutritional needs. In this volume, we discuss the dynamics of food security in communities along Canada's Atlantic and Pacific coasts, using case studies and examples to show the intricate and complex interrelationships between environmental change, economic restructuring, ecosystem health, and human health as these relate to food production and availability.

There are basic nutritional requirements for human health and well-being, as provided by diverse foods and beverages. Conversely, some of the products we consume can be harmful and impair our health, due to naturally-occurring toxins in the foods, contaminants from industrial or domestic pollutants, or from pathogenic organisms resulting from poor storage or processing methods (Committee on Food Protection 1973; Nuxalk Food and Nutrition Program 1984; Turner and Szczawinski 1991; Chan et al. 1996; van Oostdam et al. 1999; Sweeney 2003). Balancing the beneficial effects of various foods and medicines, and the potentially toxic effects of these products has long been an important endeavour (Johns 1996; Johns and Romeo 1996).

Some of the substances we consume—for example, alcoholic beverages—can be addictive and potentially harmful, yet under certain circumstances these same substances may also provide significant nutrition and, in moderation, can contribute to our health and social lives. A good example—one that links the east and west coasts of Canada—is spruce beer[1].

---

[1] A list of scientific names and authorities of the plant and animal species mentioned in this book is provided in the Appendix. Although originally made in Newfoundland from black spruce (*Picea mariana*), the same recipe could be applied to other species, as Captain George Vancouver and his ship's botanist Archibald Menzies reported. "None of that particular spruce on which they used to Brew [was] to be found near the landing place, on which I recommended another species, *Pinus Canadensis* [evidently *Tsuga heterophylla*, western hemlock], which answered equally well and made a Salubrious and palatable Beer." (Menzies' journal, July 31, 1792, near Broughton Island; Newcombe 1923). Earlier, Vancouver had travelled as midshipman on James Cook's third voyage to the Northwest Coast, so was already familiar with the brewing of spruce beer, which may have also been made from the Sitka spruce of that region.

Explorers like Captain James Cook, during his survey of Newfoundland from 1763 to 1767, adopted spruce beer as a daily beverage for his crew, evidently on the recommendation of Sir Joseph Banks, his patron, who was Director of the Royal Gardens at Kew, by appointment of King George III. Spruce beer had been made and used by Newfoundlanders since at least the 1620s.[2] Banks obtained the recipe for this beverage, the "common liquor of the country," in April 1766, on a voyage to the British Colony of Newfoundland as naturalist aboard a fisheries' protection vessel, *Niger*, in April 1766 (Justice 2000, 11).

Joseph Banks' spruce beer recipe, from his diary, September 1, 1766:

> Take a Copper [kettle] that contains 12 Gallons fill it as full of the boughs of Black Spruce as will hold pressing them down pretty tight Fill it up with water Boil it till the Rind will strip off the Spruce boughs which will waste it about one third take them out and add to the water 1 gallon of Melasses Let the whole boil till the Melasses are dissolved Take a half hogshead [a barrel holding 27 gallons], and put in 19 Gallons of water and fill it up with the Essence, work it with Barm [yeast] or Beer grounds and in less than a week it is fit to Drink . . (Lysaght 1971, 139–40).

As Jacques Cartier had discovered nearly two centuries earlier[3], Banks soon learned about the value of temperate-climate coniferous trees for antiscorbutic beverages, to prevent or alleviate scurvy, a dreaded disease caused by vitamin C deficiency, which can and often did lead to death. Spruce beer, easily made from the foliage of local conifers, was more readily available and cheaper than limes or lemons, the other well-known antiscorbutics of British seamen. Captain Cook served spruce beer routinely to his crews, and consequently, they never suffered from scurvy, which decimated the crews of other ships. When Cook arrived at Nootka Sound on the northwest coast of North America, one of his first acts was to send his crews ashore to collect spruce to replenish their supplies of spruce beer (Turner 1978).

Given its known health benefits, spruce beer would today be called a nutraceutical[4], or "health food." A number of other food and beverage products used in coastal communities could have this designation, based on recent "discoveries" of health promoting properties. Red wine, cranberry juice, wild blueberries, and partridgeberries, or lingonberries, for example, contain anthocyanidins and other compounds, which have demonstrated effects in treating visual dysfunction, impaired capillary circulation, mild inflammations, and diarrhea, and in providing retinal protection and preventing angina episodes and thrombus formation (Blumenthal et al. 2000, 16–19). Omega-3 fatty acids found in various marine oils consumed

---

[2] An entry in the book, "Golden Fleece" by Sir William Vaughan, friend and supporter of King Charles I reads: "1626 . . . Exercise best to keep off scurvy. Strong liquor prejudicial in cold countries, barley water and spruce beer best . . ." (Prowse 2002, orig. 1895: 136). Prowse (2002: 273) also documents a notation on "Beer brewed with molasses and spruce" from the Orders and Fishery Scheme of Sir Nicholas Trevanion (December 10, 1712).

[3] Jacques Cartier and his crews, over-wintering near the present site of Montreal in 1534–35, experienced severe vitamin C deficiency. They were persuaded by the local Algonquin people people to drink a tea made from eastern white-cedar boughs (*Thuja occidentalis*), and all those who drank this tea eventually recovered. A grateful Cartier described this amazing remedy in his journals, and named the tree, arbor de vitae ("tree of life"), as it is still known today (Justice 2000: 14).

[4] A term developed from combining "nutritional" and "pharmaceutical" and generally coming to mean any nutritional supplement designed for any specific clinical purpose(s). They are sold on the market as foods for general consumption (or "health foods") to be used as "supplements" to nutrition (diet).

in coastal communities are another example of a nutraceutical, which will be described later in this chapter.

Food is a prerequisite for health so not surprisingly food security is considered among the social determinants of health (McIntyre 2003). Good nutritional practices, or obtaining the required food for maintaining health, must meet the dietary needs of an individual and reduce the risk of chronic diseases, many of which are food related. Macronutrients (e.g., fat, protein, and carbohydrates) and micronutrients (e.g., vitamins and minerals) are necessary for good health. The Office of Nutrition Policy and Promotion of Health Canada has developed guidelines to assist people in choosing a healthy diet. Canada's Food Guide to Healthy Living (Canada 1992) suggests individuals consume a variety of foods from four main groups: grain products, vegetables and fruit, milk products, and meat and alternatives. The quantities of these foods recommended for consumption depend on a number of factors, such as activity level, age, body size, and gender. Additionally, it is important to recognize that not all foods are equally healthy for all individuals. Some individuals or sectors of the population may have allergy risks or intolerance to certain food groups, because of their genetic makeup. For example, many indigenous people have a low tolerance for milk and other dairy products. Diversity of dietary intake (Johns and Eyzaguirre 2003) and limiting intake of fat, sugar, alcohol, and caffeine is also recommended.

Despite the evidence supporting good nutrition, there are a number of persistent issues relating to poor nutritional practices and nutrition-related health problems. Fortunately, we are not seeing the deficiency diseases in our society, like scurvy and rickets (a condition caused by Vitamin D deficiency[5]), that we saw in previous years and well documented in the many nutritional studies that have been carried out in our study area in Newfoundland. However, we are still seeing chronic diseases that have nutritional implications, such as cardiovascular disease and diabetes; in these cases, high carbohydrate and salt intake, combined with other factors such as lack of exercise, are seen as causal agents (Friscolanti 2003b). Dental health is also a factor related to diet; excessive sugar intake, often in the form of soft drinks and other sweet beverages, can lead to tooth decay, and this problem is exacerbated by the remoteness of some coastal communities and poor access to dentists and oral hygienists.

There are other health worries over our food supplies, in the issue of toxic substances contaminating certain foods (Kuhnlein et al. 1996; Kuhnlein and Chan 2000). For example, the Centre for Indigenous Peoples' Nutrition and Environment has looked at the amount of cadmium being ingested through consumption of moose or caribou (Kim et al. 1994; Chan et al. 1997). Most of the consumption of cadmium was below the recommended level; however when cigarette smoking (contributes to cadmium in the body) was taken into account, some individuals were considered at increased risk for health problems. Many of the studies on food contaminants have been on traditional foods harvested from adjacent lands and waters, but there is a concern with a number of market foods as well because of the use of certain additives.

Many other factors interact with food to produce health problems. The level of physical activity is one of those factors. Recent concern about obesity, particularly in children, has made us think carefully about the role of diet and exercise in our lives (Friscolanti 2003a).

---

[5] The technical name is *rachitis*. Rickets, a disease especially of children, is caused by a deficiency in vitamin D that makes the bones become soft and prone to bending and structural change.

Poverty, and the lack of adequate resources to provide good, nutritious foods, is another important aspect and is one of the greatest threats to food security. In Canada three trends have been noted around poverty and food security that place poorer people at risk: (1) hunger is somewhat a hidden problem in our society; (2) money for food is compromised more than money for other essential needs; and (3) poor people are over-represented in chronic diseases related to nutrition (McIntyre 2003). The problem is often compounded in remote rural communities because food costs, especially for fresh produce, are generally significantly higher than in urban centres. Restructuring of social programs has contributed to these nutritional problems both directly and indirectly. In this volume we undertake a closer examination of issues related to food and food security, and changes that have taken place in food availability and policies related to food security for Canada's coastal communities.

To provide a more complete understanding, we place the current situation of food security for coastal communities within an historical context. The traditional and local food systems of coastal peoples have changed significantly over the past 200 years, and this change has accelerated within the past three to four decades. In general, there has been a shift away from harvesting and consumption of local food (Kuhnlein and Turner 1991; Omohundro 1994; Turner 1995) and towards the commodification of food resources in a global economy. Local communities are more and more dependent on marketed food produced outside their region, often purchased with wages from commercial food harvested or processing within their region (Nuxalk Food and Nutrition Program 1984; Kuhnlein 1984, 1992; Kuhnlein and Receveur 1996; Omohundro 1994).

Garden crops and gardening techniques were introduced by European newcomers from their earliest arrivals. For example, Table 1.1 lists fourteen vegetables planted by the Spanish on the west coast of Vancouver Island in the late 1700s. Potatoes[6], turnips, cabbage, and many other garden crops became a mainstay of peoples' diets on both coasts, and gardening was practised relatively widely in these areas until about the 1960s. At this time, changing lifestyles and attitudes resulted in less interest and time devoted to gardening. With gardens and planted orchards, even when harvesting of local indigenous products declined, people were able to supplement their diets with fresh, healthy, and relatively low-cost garden produce. Omohundro (1994, 98) described the situation for settlers in Newfoundland in the early 1800s:

> In the face of necessity, the settlers dug in . . . by 1836 there were 22 square kilometers of coastal land in vegetables and potatoes and an equal amount in pasture, a five-fold increase in productivity over that of two decades before. The burgeoning population relied heavily on home production to supplement the precious imported flour, sugar, tea and salt-beef— precious especially if fish catches were poor . . . by 1874 there were 159,000 Newfoundlanders reported to be growing an average of 6.3 bushels of potatoes each. These potatoes . . . added valuable nutrition during the four or more months of isolation imposed by sea ice.

---

[6] Potatoes, which actually originated as a domesticated crop in the mountains of Peru, Bolivia and Ecuador, were brought back to Europe from the Caribbean by early Spanish explorers, and soon became a staple in many parts of Europe, particularly in Ireland. They are still commonly known as Irish potatoes, as opposed to sweet potatoes and other tuber crops. Potatoes were brought to the Northwest Coast by the early European explorers and were quickly adopted and grown by the Haida, Tsimshian, Coast Salish and other coastal peoples. In fact, the Haida name, "*skuusiid*,"is said to be a rendering of English "good seed" (Turner 2004; Suttles 1951). The first recorded instance of potato cultivation in Canada is 1623, at Port Royal, Nova Scotia.

On both coasts, gardens were often fertilized with fish remains and seaweed. Raising chickens and livestock for meat, eggs, and milk also significantly supplemented food supplies. As these practices have diminished, access to healthy, fresh food is increasingly difficult.

Table 1.1. Vegetables grown in gardens established by the Spanish at Nootka Sound, on the west coast of Vancouver Island in 1770, as recorded by Spanish botanist Jose Moziño, under command of Alessandro Malaspina, on his visit to Nootka Sound in the summer of 1791

Barley (*Hordeum vulgare*)
pepper (*Capsicum annuum*)
potato (*Solanum tuberosum*) and several other *Solanum* species (possibly eggplant—*S. melongena*)
beetroot (*Beta vulgaris*)
carrot (*Daucus carota* ssp. *sativa*)
angelica (*Angelica archangelica*)
parsnip (*Pastinaca sativa*)
celery (*Apium graveolens*)
onion (*Allium cepa*)
rape (*Brassica napus*)
cabbage (*Brassica oleracea*)
lettuce (*Lactuca sativa*)
chick pea (*Cicer ariethinum*)
globe artichoke (*Cynara scolymus*)

The Spanish also had fenced in areas, and allowed goats, sheep, and cattle to run free.
Source: Justice 2000, 81

Today, some people in coastal communities still harvest, catch, or produce a significant proportion of their own food. However, it is recognized that food harvesting and direct food production has declined[7], and that coastal communities are increasingly relying on marketed and imported food. New roads, ferries, and commercial airlines can deliver fresh and sometimes exotic produce to these communities faster than ever before, but there is a price to be paid. Having groceries flown into remote villages represents a significant cost for delivery alone. Few can afford the luxury of such fare on an ongoing basis, and so people rely more on canned and processed goods having a long shelf life, and even these can be expensive.

Until quite recently, there have been few large-scale commercial farming operations in coastal communities. However, in Newfoundland recent efforts towards greenhouse production had the goal of supplying fresh vegetables to local and regional markets, at the same time as promoting local economic development. Some of these ventures have worked,

---

[7] Omohundro (1994) discusses the dynamics of home food production at length. One community member is quoted as saying, "People are too lazy now to do the kind of work we used to have to do [for gardening]." (p. 75). On the west coast, gardens did not grow easily, and some Gitga'at elders recalled that they were lucky if they replaced the volume of seed potatoes they were provided with by the Indian Agent. In fact, some people said, "Those seed potatoes tasted pretty good." Helen Clifton (pers. comm. to NT) said that most people had stopped trying to garden at Hartley Bay by the 1950s.

while others, which were developed largely without local consultation, have not been successful (Omohundro 1994).

Food processing and storage technologies have also evolved significantly on both coasts, sometimes beneficially and sometimes not. Because of the seasonal availability of food in temperate regions (see Figure 1.1: Seasonal Harvesting Round for the Gitga'at of the north coast of British Columbia), storing surpluses for the less productive times of year has long been essential for survival (Turner 1995, 2004). Without sufficient stored food, famines can and did occur on both coasts (Prowse 2002; Turner and Davis 1993). For many centuries— thousands of years in the case of First Peoples dehydrating and smoking were predominant means of processing food for storage. In addition, some foods, like salmon eggs, were fermented for storage, and some were stored fresh in underground caches or in boxes. The more acidic fruits, like Pacific crabapples, highbush cranberries, bakeapples (cloudberries), and partridgeberries (lingonberries), were stored fresh in containers under water, or mixed with "grease," oil rendered from the eulachon, a small type of smelt (also "ooligan" or "oulachen"; *Thaleichthys pacificus*) sometimes known as candlefish, because it contains so much oil that when dried and threaded with a wick it can be burned like a candle. All of these processing and storage methods are still practised to some extent by some families and communities. In general, these methods are safe and, if done properly, do not result in significant loss of nutrients. However, some deterioration in nutritional value is inevitable over time and there is always a danger of mould or rotting if the dehydrated food becomes damp, or if the food becomes maggoty through poor storage.

With the arrival of Europeans, the salting of fish, meat, and seafood became a major storage technology. Since the earliest days of the Basque and Portuguese fishers on the coast of Newfoundland, salt traded from the Mediterranean region was imported for preserving cod (Prowse 2002, orig. 1895). Salt cod is still a favourite with older Newfoundlanders, and especially in Portugal, where it is featured in a couple of national dishes. On the West Coast too, the Europeans introduced salt for storing food. As noted by Gitga'at elder, Helen Clifton (pers. comm. to NT, 2003), "when they introduced casks (barrels) and started salting down their food" this was a major change that started to affect their people's health. Helen recalled that even fifty or sixty years ago, using rock salt to cure and preserve their fish was a predominant method. Excessive salt intake is associated with heart disease and high blood pressure, so the health implications of this practice are potentially serious.

Freezer technology for preserving fish was first experimented with in the 1870s, but freezers require electricity, and electricity was often slow in coming to remote communities along the coasts. The Gitga'at community of Hartley Bay installed their first powerhouse and electricity generator in 1928, but it was not until the 1990s that they brought generators and freezers to their spring camp at Kiel, so large-scale freezing of their spring halibut, spring salmon, and shellfish was not an option until that time. Today, people sometimes freeze even their edible seaweed if they cannot get enough sunny days to dry it immediately (Turner and Clifton, 2006).

Food preservation by canning and vacuum bottling was developed in the early 1800s. By the late 1880s, canning technology was widespread, and canned fish and other commercially produced food started to become widely available. People in coastal and other rural communities found seasonal work in canneries that were being established all along the coast.

Winter
Trapping, harbour seal, ducks (many kinds), halibut, cockles, clams, crab, ling cod, red snapper, flounder, sea urchin, mussels, shrimp, woodfern rootstocks and other root vegetables, stored frozen or salted food.

Fall
Crabapples, highbush cranberries, lowbush cranberries, wild currants, silverweed, clover, riceroot, sockeye, chum, humpbacks (pinks), cockles, clams, sea cucumber, urchins, ducks (many kinds), harbour seal, deer, bear (grease), mountain goat, moose.

Gitga'at Seasonal Round

Spring
Oulachens & grease (Nass, Kitamaat), herring roe, harbour seal, sealion, seaweed, halibut, spring salmon, seabird eggs, mussels, chitons, abalone, cow-parsnip, fireweed shoots, thimbleberry/salmonberry sprouts, dock greens, riceroot, silverweed, clover roots, inner bark of hemlock, fir, cedar bark and roots as materials.

Summer
Huckleberries, blueberries, salalberries, salmonberries, thimbleberries, strawberries, and other berries, Labrador-tea, spring, chum, coho, sockeye, humpbacks (pinks), steelhead, trout, halibut, deer, geese.

Figure 1.1. Gitga'at Seasonal Round

In some cases, as with the clam cannery run by Gitga'at band members (Figure 1.2), they participated directly in the canning business. People also found home canning to be an efficient and practical method of preserving their harvested food, and they also used crocks and glass jars buried in moist ground to store some of their harvest for winter. Sometimes food was canned improperly and then there was spoilage and even illness and death. Even in recent times, people have fallen victim to botulism from badly processed fermented salmon eggs. Nevertheless, between freezing and canning, the opportunities for effective food storage have increased, enhancing year-round food security from local food supplies.

Figure 1.2. Clam Cannery Label Robert D. Turner collection

As noted, food quality is an important aspect of food security. It is significant not only at the human interface but throughout the food web leading to humans. Some of the critical components of our diets are synthesized at the base of the food web. As an example of how human food security and health in coastal communities is inextricably linked to the entire environment and the food web it supports, we outline the health effects of dietary seafood, including the nutritional role of "omega-3" fatty acids contained in marine animal products commonly consumed by coastal peoples.

Much has been made of the importance to human health of omega-3 fatty acids in fish oils but it is not the fish themselves that synthesize the omega-3 fatty acids. The fish have to acquire them from preceding trophic levels within the food web, ultimately starting with the phytoplankton. Phytoplankton—microscopic aquatic plant life—make fatty acids with certain omega (ω designations ($\omega3$, $\omega6$ etc.) which cannot be modified by animals, yet some of these $\omega3$ and $\omega6$ fatty acids are essential to animal growth. This is true for all components of the food web as well as for humans. The paradoxical situation resulting from this absolute requirement for these essential fatty acids is that in aquaculture we actually have to feed fish oils to fish. This commodity is one of the most expensive components of fish feed in fish farms, and demand is rapidly outpacing supply. Without these polyunsaturated fatty acids (PUFA) in the diet, fish have reduced growth rates, reduced feed efficiencies, and increased mortality (Committee on Animal Nutrition 1993). Lipids in general also serve as an important source of dietary energy for all fish, especially cold water and marine fish, which have a limited ability to use dietary carbohydrates for energy (Committee on Animal Nutrition 1993). In addition, lipids are a better source of energy for fish than proteins (Barnabé 1994).

The cold waters around Canada have traditionally been an excellent location to find marine oils. This is because these oils are used as an insulator by marine mammals, including whales, dolphins, seals, and sea lions. Before John Cabot's voyages, Basque fishers were hunting whales off the coast of Labrador (Ackman 2003), and the rendering of the blubber for the oil preceded the cod (liver) oil industry in seventeenth century Newfoundland. In the eighteenth century an industry based on the rendering of fat attached to harp seal skins developed. This oil was valued as a fuel for lamps, as cooking oil, and as a lubricant. There is now renewed interest in seal oils as a novel nutraceutical or functional food (Shahidi and Wanasundara 1998) fuelled by significant research funding, the rhetoric of politicians and media attention. The scientific basis for this interest lies in the fact that the original work on the health benefits of $\omega3$ fatty acids was done on Greenland Inuit who consumed fats from marine mammals more than they did from fish. The $\omega3$ fatty acids are located differently in the triglyceride storage molecules in marine mammals and fish. On the West Coast, "grease," or eulachon oil, mentioned earlier, which contains $\omega3$ fatty acids, is a staple (MacNair 1971; Kuhnlein et al. 1982; Drake and Wilson 1991). The oil is of great cultural significance for many First Nations peoples of the Northwest Coast, who consumed, and still do consume grease with almost every meal as a condiment or flavouring agent, as well as using it for trade. The Nuu-Chah-Nulth and other West Coast peoples also consumed substantial amounts of seal oil, whale oil, and salmon oil (Turner et al. 1983; Drucker 1951).

Cold oceanic temperatures also ensure that marine lipids in organisms have a high degree of polyunsaturation. Lipids are critical components of the biological membranes surrounding animal and plant cells, and in order to remain fluid at cold temperatures these membranes need unsaturated fatty acids. Almost all PUFA are of the $\omega3$ and $\omega6$ type. In addition to having structural roles in membranes, some of these ω3 and ω6 PUFA are also precursors of highly biologically active molecules once known as "local hormones." Thus, all steps in the food web leading to humans require ω3 and ω6 PUFA for both biophysical and biochemical reasons.

For humans, the positive health effects of dietary fish oils are largely related to the ω3 PUFA content. Consumption of ω3 fatty acids is known to have beneficial effects on plasma lipids and lipoproteins, reducing risk of cardiovascular disease, cancer, adipose tissue mass, and inflammatory diseases. It is well established that dietary fish oils affect the amounts and types of fats in the blood stream. Omega-3 fatty acids consistently lower circulating levels of blood triglycerides in humans. Use of "practical" doses of ω3 acids in humans has a significant and probably clinically important effect on serum triglyceride concentrations, especially in hypertriglyceridemic patients (Harris 1997). Reduction in plasma triglyceride concentrations was one factor that early studies suggested might improve human health by reducing atherosclerosis. More recently, the preferred mechanism for reduction of cardiovascular mortality is a lowering in the heart's susceptibility to arrhythmia (Hallaq and Leaf 1992; Goodnight 1996), making it less prone to the irregularities thought to be a lethal element in a heart attack.

In addition to heart disease, dietary fat intake has been linked to cancer. Several studies have suggested that ω6 PUFA act as tumour promoters (Tisdale 1992) and that ω3 fats reduce the risk. A perhaps unexpected finding is that dietary fish oils can also limit fat cell growth in rats (Parrish et al. 1989, 1990). The decreased fat accumulation has led to the suggestion that fish oils be used in the treatment of obesity (Groscolas 1993). The smaller mass of fat also helps reduce the concentration of lipophilic organochlorines in the body as it limits storage in adipose tissue and consequently increases metabolism in the liver (Umegaki et al. 1995).

Omega-3 PUFA can also play a role in controlling the body's protective inflammatory processes. Inflammation is a defensive reaction of the body to repair damage and, if it falls out of equilibrium, it can cause problems such as rheumatoid arthritis and psoriasis. Fish oil therapy has also been suggested for treatment of Crohn's disease (O'Sullivan and O'Morain 1998), otherwise known as regional ileitis, and for nephropathy (Grande and Donadio 1998), or kidney disease.

The complex relationships between human health and nutrition and the intricacies of the marine food web are exemplified by this story of ω3 fatty acids. There are many other stories to be told that link human well-being and food production. There are major, ongoing, and escalating concerns in terms of access and availability of good quality local food by people in coastal communities on both coasts. The two major issues here are diminishing food resources, and restriction of access to available food resources. Diminishing resources—especially in the fisheries and shellfish—have been noted since the first commercial scale exploitations of fisheries and other resources. For example, the great auk, which used to inhabit the coasts and offshore islands of Newfoundland and Labrador were hunted to extinction by 1850 (Montevecchi and Tuck 1987). A whole range of fisheries, from pilchards to whales, and including cod and salmon, abalone and lobster, have experienced boom and bust cycles, as these resources have been almost universally overexploited and improperly managed. These and other food species have also been impacted by habitat deterioration and loss caused by a wide range of factors: poor forestry practices that destroy spawning grounds; bark and wood depositions on the sea floor from log booming; dredging and breakwater construction; industrial and domestic pollution; and invasive species incursions (Ommer and team, in press).

At the same time, coastal communities' access to the productive land base that supports local food production has been increasingly restricted. This process started for the First Nations upon the arrival of the first European colonists, and continued with the establishment of reserves and the confining of aboriginal communities to them (Harris 2005). Privatization of the land-base, or placing it, through forest tenure leases, into the control of corporate interests, as well as creation of parks and other protected areas, has further eroded local peoples' access to and control of their traditional lands, waters, and resources (Lepofsky et al. 1985; Turner 2005; Ommer and team, in press; Turner 2005).

Ultimately, the reduced availability of traditional food combined with lifestyle changes, such as participation in the wage economy, has resulted in a reduction of knowledge and experience about traditional food and how to harvest and prepare it. Ironically, this trend has also resulted in a reduced appreciation of traditional food among people who have had little chance to experience it, as reflected in interviews with three generations of Nuxalk women (Kuhnlein 1989, 1992). These factors all have direct implications for health in coastal communities (Stephenson et al. 1995). The social aspects of food systems are also a critical element of health (Wilson 2003; Ommer and Turner 2004) and are affected by the declining use of traditional food: division of labour, the role of clan systems in food production, importance of a wide range of feasts in which social knowledge and practices are conveyed along with the food being served, and the passing on of many kinds of practical and social knowledge between elders and youth as food is being procured and processed.

Today, people in many coastal communities experience unacceptably low food security, as evidenced by the need to use food stamps, by poor nutrition amongst school children in some cases, and by very real health problems, both emotional and physical that people in many coastal communities are reporting. The old adage, "we are what we eat," is good to remember as we consider the effects of these dietary changes. There are many important links between environment, lifestyles, economies, social structures and institutions, knowledge acquisition and transmission, biodiversity conservation and human nutrition, and food security. In the following paragraphs, we provide a framework for the rest of the volume, contextualizing the major sections and their chapters within the overall topic of food security in coastal communities.

Section 1 of the book provides a more in-depth coverage of the biophysical aspects of food—particularly marine food—and connects these to human food security. It contains chapters that deal with food webs and fish-fisheries dynamics, with a discussion of the role of theoretical modeling (Chapter 2: Martinez Murillo and Haedrich); and nearshore marine food webs and human impacts, focusing on three specific case communities, one in Labrador, one in Newfoundland and one in British Columbia (Chapter 3: Parrish, Copeman, Van Biesen and Wroblewski). Food-web modelling is also a topic covered in Chapter 4, which discusses requirements for restoring marine ecosystems to some level of their past abundance, biodiversity, and trophic structure, and the policy implications for maintaining food security and resilience of coastal communities (Haggan, Ainsworth, Pitcher, Sumaila, and Heymans). Aquaculture practices and their effects on ecological integrity and food security are the focus of Chapter 5 (Volpe). The section concludes with a discussion on changing resources in coastal Labrador (Chapter 6: Kennedy).

Section 2 addresses the dynamics of livelihoods and food production as they relate to coastal communities and their history. Chapter 7 provides case examples of past and present hunting practices on the Atlantic coast and relationships with restructuring (Montevecchi, Chaffey, and Burke). It also discusses the relationship between food security and legislation on Canada's coasts, looking at some of the historical developments and their effects. Chapter 8 looks at informal economies and subsistence strategies that have helped to sustain coastal communities, and their contributions to resilience (Ommer, Turner, MacDonald, and Sinclair). It also examines the history of commercial food production and its role in environmental and social restructuring.

Section 3 covers the cultural aspects of food production, food use and food security. Chapter 9 discusses the relationship between public school education, nutrition, and health (Harris and Shepherd). Chapter 10 links food with cultural identity and social health, providing case examples of the ongoing importance of food in feasts, ceremonies, and celebrations (Tirone, Shepard, Turner, Jackson, Marshall, and Donovan). In Chapter 11, the relationships between cultural health and social health and economic restructuring are emphasized (Solberg, Canning, and Buehler).

Finally, in the concluding section, *Section 4*, we focus on the regulation of food production and food access and on the policy implications for improving food security for people in Canada's coastal communities. Chapter 12 focuses on health and nutrition and the occurrence of dietary deficiency diseases in the past century in Newfoundland (Kealey). Chapter 13, the final chapter, provides direction for reassessing policy and regulation with a view towards developing greater environmental and social sustainability for Canada's east and west coasts and towards increasing food security and resiliency in our coastal communities (Parrish, Turner, Ommer, and Solberg).

ACKNOWLEDGEMENTS

Thanks for their contributions to: Chief Johnny and Helen Clifton and their families; Belle Eaton, Clyde Ridley, Colleen and Gideon Robinson, Tina Robinson; Ken Campbell, Dr. Margaret Seguin Anderson, Judy Thompson (Edosdi), and Dr. Robin J. Hood.

In: Resetting the Kitchen Table
Editors: C. C. Parrish et al., pp. 13-32

ISBN 1-60021-236-0
© 2008 Nova Science Publishers, Inc.

Chapter 2

A PERSPECTIVE ON NEWFOUNDLAND'S FISHERIES ECOSYSTEM USING SIZE-BASED FOOD-WEB RELATIONSHIPS

María de las Nieves Martínez Murillo and Richard L. Haedrich

Department of Biology, Memorial University of Newfoundland, Canada

ABSTRACT

We present a conceptual model wherein fish and fisheries interact in an action–reaction feedback loop, with a unidirectional flow of biomass to the fishery. The fishery persistently targets one or a few species until the fish community balance breaks down and the system's structure changes when the target species becomes scarce. The response of the fish community reflects back onto the fishery, which diversifies and chooses new target species according to availability, abundance and economic value. Nonetheless, the fish community defined by its populations of interacting species sets the biological limits within which the fishery takes place.

An hypothesis of size structuring the fish community is central to our chapter. Size is related to species life cycle and those species interactions, especially predation, that determine the structure of a fish community and drive its dynamics. Fisheries also act on size by targeting the larger individuals. Therefore, size structure in the fish community can reflect and be a direct result of fish–fisheries interactions.

Using empirical data from the fish community off Newfoundland, we examine the implication of size at the different levels of organization and analyse how the size-structured community changes as a result of strong disturbances such as an aggressive fishery. We argue that size is key in determining the structure of the community and its dynamics, and argue for a quantitative size-based approach to simulate the natural dynamics of a fish community and how they might respond to stress from externalities.

In this chapter, we build the case for a novel size-based simulation approach and present a quantitative model based on the Newfoundland system. This model is stable over runs of centuries and can serve as a framework to examine the possible consequences under different management scenarios. Preliminary findings are that: (1) recovery to stable conditions after a disturbance can take hundreds of years, (2) the system recovers to previous conditions after an environmental disturbance, (3) the system may return to a less productive state after a fishing disturbance, and (4) the state of recovery after a combination of environmental and fishing disturbances is unpredictable.

ON THE MATTER OF FISH–FISHERIES INTERACTIONS

Introduction

We start from the premise that fisheries are one of the world's most important food sources but that our fundamental concern is with their ecology. As a part of *Coasts Under Stress* we studied fish size at different levels of organization and analysed how the size-structured fish community[1] changes under fishery disturbance. This chapter presents the basis for the larger and more technical work of the resulting PhD thesis (Martinez Murillo 2003). We argue that the size factor is key in determining the structure of the community and its dynamics, and have developed a size-based approach and model to simulate the natural dynamics of a fish community and how those dynamics might be altered in the face of stress from external forces.

No ecosystem in the world escapes human action (Vitousek et al. 1997). Humans to some extent influence all earth's ecosystems, often through direct exploitation. Such is the case of fisheries, which are an important source of protein for humans. When the ecosystem has the capacity to regenerate resources cyclically, these resources are called "renewable" and fisheries are a prime example.

However, the status of renewability is not always maintained and can change in ecosystems under intense or prolonged stress (Odum 1985; Rapport et al. 1985; Rapport and Whitford 1999). Depletion of resources occurs when the rate at which a resource is taken exceeds the rate at which that resource is naturally produced, or when the activity of taking has a negative indirect effect on natural production.

Most usually, the rate at which resources are exploited increases in response to social development (Deimling and Liss 1994; Ommer 2002a). Resource exploitation creates jobs and economic profits. If the activity can support many social and economic activities, the exploitation also becomes important as a political and economic tool for any government involved. The goal moves from subsistence to profits and the measurement unit is no longer the resource itself but the money obtained from its exploitation. With the short-term view to maximize benefits and profits, more effort and more people enter the business of exploitation not considering or unaware of the risks of overexploitation. The risk is even greater when there is a competition for limited resources and everyone wants to make the most of them, a situation known as the Tragedy of the Commons (Hardin 1968).

This appears to apply to many fisheries. Regulations for exploitation cannot easily be imposed when several countries are involved and agreement is difficult to reach. What usually happens is that sooner or later exploitation exhausts the ecosystem's productive capacity, at which point the system cannot respond to human demands any longer. This triggers chaos at natural, social, economic, and political levels (Sherman 1994; Harris 1998; Haedrich and Hamilton 2000). On the one hand, economic profits stop or become losses. The

---

[1] A community is a group of interacting organisms that inhabit a given area. Communities are made up of populations of different species (see Smith and Smith 1998).

immediate consequence is that people lose their jobs; they need an immediate solution because they need to make a living. On the other hand, resources may have disappeared for good or may need a long time to recover.

From the moment the depletion becomes serious the only choice humans have is to either permit the resources to recover (if it is still possible) or to keep exploiting whatever is still left. The conclusion is that humans may have power over resources in the short term, but the resources ultimately set the limits for their own exploitation (Folke et al. 1993; Daily 1997), and hence the importance of understanding the natural dynamics of the resources in order to avoid the serious social consequences of their depletion.

For a long time, fish in the sea were considered inexhaustible (Smith 1994) or treated as though they were limitless by newly developing or expanding fisheries (Merrett and Haedrich 1997). However, fish communities worldwide are suffering drastic changes. "Why are they changing?" and "How are they changing?" are questions still in debate, and now another one has been added: "Will they recover?" The answer begins with an ecological consideration of the fish community, but quickly moves to social consideration regarding the human community.

The future of fisheries resources is no longer the exclusive concern of fish biologists. "Sustainability" is now the focus of ecosystem management (Sherman 1994; Olver et al. 1995; García 1997; Zabel et al. 2003; Pikitch et al. 2004). There is need for a clear definition of this term (Hueting and Reijndiers 1998; Pendry 1998; Svirezhev 1998; Haedrich and Hamilton 2000; Phillis 2001; Ayres et al. 2001) and for consensus among the diverse social sectors about sustainability of what, for whom, and for how long. Sustainability at economic levels may be incompatible with sustainability at ecological levels (i.e., for the resources and their habitats) unless an objective measure is used to weigh resources versus profits. Constanza *et al.* (1997) estimated the economic value of the resources provided free by nature to be something like US$33 trillion per year, contrasting with the global gross national product of US$18 trillion per year.

The inclusion of humans as part of the ecosystem (Barrett 1985; Stephenson and Lane 1995; Newell and Ommer 1999; Coward et al. 2000) is a step towards the unification of the term sustainability, and is probably the most convincing argument for rigorously advocating ecosystem conservation. Simply stated, we are all in this together. Nonetheless it seems as though the resources instead of the humans have been neglected, to judge from the imperilled state of many ecosystems as a result of human activities. It is now clear that future fisheries management will require putting sustainability, ecosystem health, and human welfare together in the same context, as has been recommended in the US (National Research Council 1999).

History Repeats Itself in Fisheries

The general pattern of the interaction between fish and fisheries can be described as an action–reaction feedback loop (Figure 2.1), with a unidirectional flux of biomass towards the fishery (Regier and Loftus 1972; Jackson et al. 2001, Pauly and Palomares 2001). Within this framework, humans, acting as a selective predator, target one or a few species (oblique arrows). They proceed intensely and persistently (down arrows) until the community balance breaks down and the system's structure changes (lower right arrows). This change becomes evident when the target species becomes scarce. This response of the fish community reflects back upon the fishery, which changes (upper right arrows) by selecting new target species mainly as a function of availability, abundance and economic value. This causes a new perturbation and the process repeats itself.

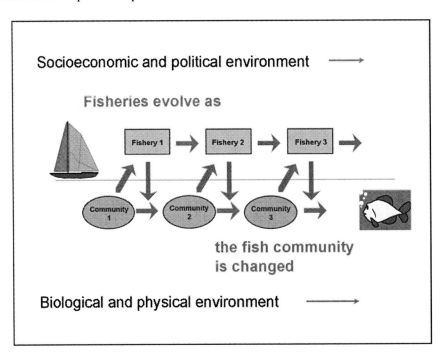

Figure 2.1. A conceptual model of fisheries and fish community dynamics (see text)

In Figure 2.1 the starting point is the unexploited fish community (Community 1). Any fishery has to act on some existing resources. Identification of this community can indicate the kind of fishery it may support and the probable resultant trend of that fishery over time. Furthermore, as indicated in the top and bottom of the figure, the naturally-occurring community of fishes and the fishery (the human community) are each situated in their own separate framework of environmental and socio-economic changes that no doubt develop and evolve at rather different rates.

The long-term direction of the interaction loop is alarming (Ludwig et al. 1993; Smith 1994; Weber 1994; Safina 1995; Buckworth 1998; FAO 2000). Fisheries all over the world are characterized by an initially prosperous and fast development followed by fishery collapse (Hilborn and Walters 1992). The main long-term changes in the community of fishes are: (1) the age composition of target species changes towards younger, smaller-sized, individuals (North Sea Task Force 1993; Large et al. 1998; Bianchi et al. 2000), and (2) catch composition shifts from larger, long-lived, top predators down to smaller, short-lived, lower trophic level species (Merrett and Haedrich 1997; Caddy and Rodhouse 1998; Pauly et al. 2001).

Figure 2.2 shows the world landings of marine fishes since 1950 (from the FAO database program "FISHSTAT+"). Over forty years, landings by developed countries exceeded by far that of developing countries. The difference in landings is especially remarkable during the 1950s, a time in which fisheries production in developing countries was very low. Large-scale fisheries in developed countries had started sooner, and their peak and then decline in the

1990s corresponds with resource collapses followed by the application of more strict rules concerning exploitation.

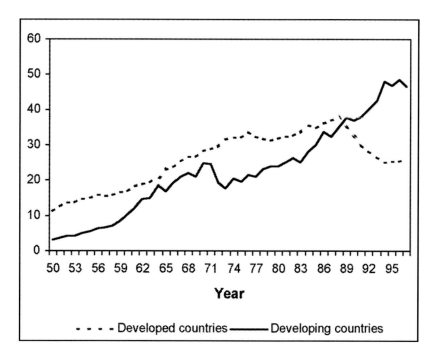

Figure 2.2. World production of marine fish, from FAO "FishStat+"

Meanwhile, landings in developing countries increased rapidly, probably due to economic needs but also linked to more relaxed attitudes as regards management for sustainability. The trend in developing countries since 1990 shows a steeper increase than ever followed previously either by developing or developed countries. Thus developing countries will reach their peak production in a shorter time than was experienced by the developed countries. Exploitation of marine resources other than fish is also accelerating in developing countries, so full ecosystem effects are almost certainly taking place.

Other features related to fishing activity also show the trend of the world fishery. According to FAO the number of fisheries has doubled since 1970 and in 1990 28.5 million people were involved. The increase is uneven, taking place mainly in Asia while the number of fishers decreases in industrialized countries. The same tendency shows up in the number of vessels. The fishing fleet is decreasing in developed countries and increasing in developing countries. The same scenario that lead to a fishery collapse in developed countries is now being replayed in the developing world.

The history of fisheries for the Northwest Atlantic dates back to before the official discovery of North America by the Europeans, and has been reported recently by, e.g. Hutchings and Myers 1994, Kurlansky 1997, and Lear 1998. The following summary draws on these publications.

Right after 1497 the Portuguese, French, and Spanish started fishing for cod in Newfoundland waters and were followed by the English in the mid-sixteenth century. All supplied mainly salted dry cod to European markets. The initially seasonal fishery with boats sailing from Europe in the spring–summer time expanded with the settlement of the territories by the end of the sixteenth century. There were signs of overexploitation as early as 1713, when declining inshore catch rates prompted the expansion of the offshore banks fishery. At that time a small cod could weigh sixty pounds and a large one a hundred pounds, much larger than any seen today. During the eighteenth century fish processing became the industry and way of life all along the northwest Atlantic coast. With fluctuations in the cod fishery other species became important during the nineteenth century, including Atlantic halibut (*Hippoglossus*), haddock (*Melanogrammus*), and flounders (Pleuronectidae). Drastic changes in the fishing methods appeared in the twentieth century. Inshore fishery boats started to use gasoline engines, draggers and otter trawlers were introduced in the offshore fishery, and refrigeration became available. Fish could be captured more rapidly and vessels could spend longer periods at sea. Catches increased rapidly and fish products were used both for food and industrial purposes. Groundfish catches peaked in the 1960s and have decreased significantly since then. In 1992 a moratorium on cod fishing was established and to date it is still in place with no sign of recovery. In 2003, the Committee on the Status of Endangered Wildlife in Canada (COSEWIC) determined that the Atlantic cod was an endangered species.

The second half of the twentieth century has seen the collapse of the centuries old fishery off Newfoundland. Crucial for this situation has been the fisheries intensification of the 1960s and the lack of strong recovery afterwards (Hutchings and Myers 1994; Murawski et al. 1997). The total groundfish landings went from one million tonnes in the 1950s to two million tonnes in 1965. Cod catches more than doubled, reaching a maximum of 810,000 tonnes in 1968. In 1977 Canada and the United States extended their fisheries jurisdiction to 200 miles. This stopped the foreign fishing, but expansion of the domestic fishery quickly filled the gap left by the foreign fleets and resources were driven to collapse by 1990.

Studies on fisheries resources, which started more than a century ago (Megrey and Wespestad 1988; Smith 1994), have not succeeded in preventing collapses (Botsford et al. 1997; Longhurst 1999). The faster development of fisheries in comparison to scientific understanding of their basis (Haedrich et al. 2001) contributes to this failure. In addition, the emphasis on research for fisheries development favours a short-term view of the problem and its solution, which is a major handicap as regards sustainability. Perhaps even more importantly, any human control of the dynamics of widespread marine fish resources is limited. Interacting forces and time lags influence marine fish community dynamics to the point where the word that best reflects fisheries in the long-term is "uncertainty."

Does all this make management or sustainability a utopia (Ludwig et al. 1993)? As a part of ecosystems, humans are at most times a dominant player and the one capable of more flexible and purpose-oriented actions (Gislason et al. 2000). If fisheries are no exception, it should be possible to regulate the state of natural fish communities to a great extent by regulating human activities. There are uncertainties concerning fish community dynamics that humans cannot control. Therefore, it seems sensible to talk about management and sustainability in terms of the way humans interact with the ecosystem. At the same time, it is clear that understanding how fish communities operate is crucial for fisheries management.

Uncertainty in the Dynamics of Fish Communities

There are three sources of uncertainty in the study of fish dynamics; environmental, social, and biological forces all act and interact in the dynamics of fish communities. Within the conceptual model depicted in Figure 2.1 social forces are indicated at the top of the figure where they drive changes in the fishery, and environmental forces are indicated at the bottom where they drive changes in the fish community.

The environment sets the basic conditions for the wax and wane of natural populations. Environmentally driven population fluctuations prior to fisheries development have been described (MacCall 1985). But today there is also strong evidence that fisheries may cause the collapse of fish populations (Hutchings and Myers 1994; Myers et al. 1996; Walters and Maguire 1996). Government policies and the economic market are a second source of uncertainty because they change the intensity of fisheries and the way fish resources are exploited, thereby adding variability to the dynamics of fish communities. Finally, our incomplete knowledge of the structure and interactions within the fish community itself also creates indeterminacy (Gomes 1993).

While environmental uncertainty is uncontrollable, human activities usually can be regulated, making them a source of variability rather than uncertainty. Biological uncertainty lies within limits constrained by the life history parameters of the species and is reduced when considered within the structure of and interactions in the community of fishes (Ulltang 1996). Furthermore, the relative importance of environmental, human, and biological factors is not always equal. In stressed ecosystems the importance of biological processes is heightened. While fish have adapted to environmental variability through mechanisms such as varying life histories, fisheries development has been so fast that fish have not been able to adapt to this new pressure. At that point, density-dependent biological processes become predominant forces in the dynamics of the fish community. Seal predation, for example, could be a problem for cod once fish populations have been reduced to very low numbers by overly aggressive fishing.

Most commercially important fish species spawn large numbers of progeny. Yet the unpredictability in environmental variation makes recruitment a hazardous process because survival of the eggs and larvae is highly dependent upon the environment in which they develop. Nonetheless, an evolutionary adaptation to accommodate a natural range of environmental changes has certainly occurred in all species. Recruitment is also a density-dependent process (Bjorkstedt 2000; Myers 2001) influenced by spawning stock size and predation. When fish are abundant the effect of density-dependent processes is less evident than that of environmental effects. However, at low levels of abundance these density-dependent processes may have a greater relative effect than recruitment. Hence, fish community structure and interactions, involved in density-dependent processes, are important factors to consider in the study of dynamics in a community of fishes.

Predation, human activities including fisheries and environmental factors are the main causes of mortality in fish species. As a biological factor, predation is intrinsic to the community and at some level acts regardless of the presence of the other external environmental or human factors. Moreover, predation is the link among individuals in the community, determining the indirect effects of external factors, or externalities, on the whole community. Defining the structure and interactions in a fish community will assist in the study of the effects of externalities on the dynamics of the fish community. Body size appears to be a good indicator of the structure and interactions of fish communities.

The Importance of Size

Fisheries research and management would profit from the refreshing views of J. T. Bonner (1965). In the two first chapters of his book *Size and Cycle* he challenges the reader to think of organisms as life cycles and to use size, as a characteristic of this life cycle, to make comparisons among organisms. In the marine environment, juveniles of different fish species are ecologically more similar to each other than are juveniles and adults of the same species. Habitat and diet, for example, are usually shared by individuals of different species when at the same developmental stage, but not by individuals of the same species in different stages. If we consider characteristics such as food requirements and behaviour, it should be possible to group fish individuals according to their life cycle stage and habitat requirements regardless of the species to which they belong. Hastings (1988) argues for cases where age or size structure must be used when studying population dynamics. These include competition in juvenile and/or mature stages, cannibalism, dispersion, and predation, all of which are matters of concern for understanding the dynamics of fish populations.

The importance of size at the different levels of organization, i.e. from individual to ecosystem, was recognized early in the history of modern ecology (Elton 1927). Treatises on size and allometric rules[2] concentrate in the mid 1980s (McMahon and Bonner 1983; Peters 1983; Schmidt-Nielsen 1984; Calder 1984). Although it is difficult to find general laws valid for all ecosystems and organisms, constraints that the marine environment imposes on marine organisms make it likely that allometric rules apply for most organisms and levels of organization there.

Many biological and ecological characteristics of species are related to size (Peters 1983; Calder 1984). Allometric rules generally govern physiological processes at the individual level; e.g. metabolic rates of small juveniles are often greater than those of older and larger individuals. At the species and population level, allometric rules applied to life-history strategies allocate species along the r-K spectrum; e.g. r-selected herring (*Clupea*) broadcast large numbers of small eggs and the more K-selected wolffish (*Anarhichas*) lay large eggs in rocky nests. At the community and ecosystem level, allometric rules explain the way a species utilizes its environment, for example in regard to geographical distribution, prey selectivity, and trophic relationships; small fish species (e.g. Atlantic herring, *Clupea harengus*) are often regionally distributed plantivores and large fish species (e.g. bluefin tuna, *Thunnus thynnus*) are more widely distributed piscivores.

The size factor that underlies allometric rules is of special relevance among fish species. Most fish start their lives as very small organisms and grow steadily over their entire lifespan (Woodhead 1979) making for a strong correspondence between size and age, and the energetic requirements of an individual are in accordance with its growth (Peters 1983). Maximum size of individuals in a population is related to lifespan (Calder 1984), and lifespan (and in particular reproductive age) determines the time a population needs to adjust to local disturbances. Fecundity is also related to size (Wootton 1979). At the community level, feeding relationships occur between individuals rather than species, and trophic interaction is also based on size. "Big eats small" (Hahm and Langton 1984; Lundvall et al. 1999) is the norm, with fish being mostly opportunistic feeders (Lilly 1987, 1991, 1994). From this

---

[2] Mathematical expressions relating organism size to physiological, life history and ecological parameters. Allometry refers generally to changes that occur differentially as a consequence of growth.

perspective, allometric rules play a significant role from the individual through to the ecosystem level. Even the spatial distributions of individual fish change over their lifespan in relation to their size.

The general rule that big eats small, the reality of trophic interactions as the main links among species, the fact that predation constitutes the main cause of fish natural mortality, and the dominance of opportunism within the feeding habits of fish species all support the idea that size is a key factor in determining the structure of the fish community. This hypothesis of size structuring the fish community is central to our thinking about fish community dynamics. Due to fish characteristics, allometric rules in fish transcend the individual level to the population and community levels. Size is related to species life cycle and species interactions, which in turn determine the structure of a fish community and drive its dynamics. In addition, fisheries usually act to influence size structure through targeting the larger individuals. Therefore, size structure is very likely to reflect fish–fisheries interactions and thus constitutes for us a unifying metric (Dickie et al. 1987; Pauly et al. 2001).

Temporal and Spatial Scales

A glance at any historical set of landings or survey data shows that the answer to the question, "how is the fishery doing?" depends on the point at which we ask the question and how wide a scope we want to examine (Haury et al. 1978; Post et al. 2002). A continuous increase in fish abundance over a decade may be of little importance to a temporal pattern on the scale of a century. Similarly, an increase of fish at a certain location may be insignificant in the context of a bigger area. Scale to be considered must be in accordance with the problem one wants to address. It will not be possible, for example, to evaluate the state of Atlantic salmon (*Salmo salar*) in North America if the study is limited to one river or lake. Neither will research conducted for a period much shorter than the species' generation time, or covering only a single spawning hotspot, inform much about the state of an entire fish population.

The immediate effect of a fishery on a fish community, in the absence of any mechanisms for enhancement, is the reduction of abundance of one or several species. This impact is effectively instantaneous, however it is only after a longer time that the full impact of the fishery shows up; there is a time lag until inter- and intra-specific interactions manifest themselves. The reaction of the fish community depends on species generation times and on individual interactions. Thus, the immediate fishery impact can be amplified or buffered by relationships within the community. Amplification can occur, for example, when fisheries removal of one species (e.g. large brook trout, *Salvelinus*) favours the increase of a competitor (e.g. common suckers, *Catostomus*) as we have observed in lakes of southern Labrador. In contrast, a buffer effect can lead to a reduction in predation on young individuals when cannibalistic species like Atlantic cod are fished. Many of the long-term effects of fisheries spread through the community as a result of such trophic interactions (Vanni et al. 1990; Parsons 1992).

A good criterion to use when choosing appropriate scales is to choose a scale sufficiently large to allow observation of the full range of variation in what is being studied (Powell 1989). Nonetheless, the longer the time period we consider, the better we can interpret changes observed in a fish community (Jackson et al. 2001). Due to the long lifespan of many fish species, a long-term period is required to assess the trend of how a species reacts to a fishery (Connell and Sousa 1983), particularly when fisheries target a certain stage (size) of the species, usually the largest individuals. Removal of large mature individuals, among other effects, will reduce the number of offspring in future generations. However, the already existing younger cohorts of the species may not be affected by removal of larger individuals or may even increase in number through a reduction of cannibalism. The already existing cohorts will replace the removed individuals and mask the effect of their removal until newborn generations reach adulthood. This is why a time very much longer than a generation may be required to fully appreciate the results of inter-specific interactions (Jackson et al. 2001).

In addition there are changes in the horizontal and vertical distribution of a fish species during its lifespan. The appropriate spatial scale to look at the dynamics of fish communities should account for these changes in relative distribution as well as for the overall geographic range in which species co-occur. In many cases the spatial distribution of species is related to the presence of other species, e.g. Atlantic salmon and the introduced brown trout (*Salmo trutta*) differentially displace one another depending on local stream conditions in Newfoundland (Gibson and Haedrich 1988). Therefore, consideration of species that co-occur helps to find an optimum spatial scale to study species interactions, and thus community structure and dynamics. The spatial scale of fish communities can be of the order of hundreds of kilometres (Gomes 1993). Contemporary fishery research increasingly tends towards an ecosystem scale approach that is likely to include representatives of most species and most life stages present. At this scale of community consideration general patterns, which would not be evident considering single populations or species, can emerge (Maurer 1999). In addition, both environmental and fishery processes operate on this large scale and therefore allow an ecosystem perspective (Sherman 1994; Mann and Lazier 1996; Haedrich 1997).

APPROACHING FISH COMMUNITY DYNAMICS

The Size-Based Fish Community

In the first section we considered the context in which fish–fisheries interactions take place. Within that context, attention was directed towards the fish community for two reasons: (1) the resource must be able to sustain exploitation, and (2) environmental and human disturbances can be considered externalities whose final effect depends on community dynamics. We now turn to the fish community we have studied in the Northwest Atlantic. The study area comprises the continental shelf off the coast of Newfoundland and Labrador, and we refer to the organisms and their habitat in this area as the continental shelf fishery ecosystem.

The largest source of data available concerning this ecosystem corresponds to the demersal (bottom-living) fish, and that component is our focus. We define that group as a community to emphasize the existence of interactions among species (Gomes 1993; Paine 1994), but so doing has a twofold implication that must be supported.

On the one hand, treating all the demersal fish as a unit means adopting a community approach explicitly. This represents a departure from the norm in that fisheries research has for a long time neglected biological interactions and focused on single species populations. But the conduct of marine fisheries research itself presents additional difficulties (Mitchell 1982; Paine 1984; Steele 1984; Smith 1994): the deepwater oceanic habitat does not allow direct observation, surveys are expensive, and reliability of data obtained from landings is questionable (e.g. because of bycatch, discards, unreported catches, and similar data-fouling problems; Metuzals et al., submitted). Nonetheless, even now, most approaches consider only single populations despite the fact that biological knowledge of species is currently broad, that fisheries have diversified extensively, and that collapses have emphasized the importance of species interactions (Pauly 1988; Dugan and Davis 1993; Orensanz et al. 1998). A single-species approach has proven to be limited (Beddington 1984; Beverton et al. 1984; Larkin 1996; Botsford, et al. 1997), especially in regard to the long-term effects of externalities. As early as the end of the nineteenth century, Lankester (Smith 1994) pointed out the importance of species interactions in fish stock changes. The strong suggestion was that other species should at least be considered even when the focus was on the dynamics of just one species.

Any individual's life is constrained into a certain time period, the time between birth and death. Life span is determined genetically and can be modified by the environment, but it is always time limited. Only at the next level of organization, the population level, is there the possibility of continuity over time. Whether a population persists over time or not depends on birth and death rates. A population will have continuity over time as long as birth rate is equal to or exceeds death rate, and thus reproductive capacity becomes important. But also important in determining community dynamics are the interactions between species. Therefore the permanency of a species has to be considered at the community level. If trophic interactions take place between organisms of different size rather than between specific species, then we can talk about continuity of the community regardless of the permanency of any individual species. Thus all species within a habitat together form a unit, the community.

On the other hand, if we only consider a certain number of species as components of the community, this group of species should present features that differentiate them from other components of the ecosystem apart from the practical reason of data availability. In the deepwater marine environment, plankton, invertebrates, fish, marine mammals, birds and humans are linked by trophic interactions. However, there are some characteristics that allow us to both differentiate demersal fish species from other taxa and to treat them as a unit. The temporal and spatial scales at which these fish live their lives (Steele 1978), as well their physiological and behavioural characteristics, serve to distinguish them; e.g. fish generally range more widely than most invertebrates and their reproductive strategy is distinct from that of mammals or birds.

Size implications inherent in these characteristics create a common ground for study as a community. Using size, individuals can be classified regardless of the species they belong to, allowing treatment of the community as a whole instead of as a cluster of separate species. Strong interactions among fish species in comparison to trophic links with other species of the ecosystem also serve to justify our focus. Bax (1991) showed that predation of fish by other fish is a far more important cause of mortality than is predation by mammals, birds or humans, and Jennings et al. (2002) observed that competition for food between fish and invertebrates is low. Since predation is the most important link among fish individuals, to deal only with fish reduces the influence of other non-trophic interactions that are probably negligible.

The Nature of the Available Data

Lack of long-term scientific data is one of the main difficulties in studying fish community structure over time and space. Our work focuses on the continental shelf within Canada's 200-mile limit and NAFO subdivisions 2J3K, a deep shelf area of around 270,000 square kilometres (Figure 2.3). We consider two kinds of data: fisheries landings data from 1960 to 1994, and scientific survey data from 1978 to 1993. The fisheries and scientific survey datasets complement each other in the study of the demersal fish community. What are the advantages and shortcomings of each?

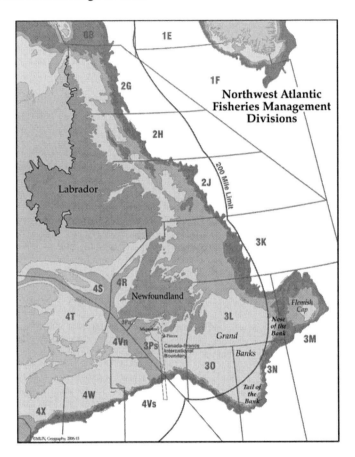

Figure 2.3. Study area, the playground for Newfoundland's fishery: NAFO areas 2J3K inside the 200-mile limit at the centre of this chart

Fishery Landings Data

Fishery data do not pretend to reflect the state of an entire fish community. If anything, fishery data reflect trends rather than accurate estimations of fish abundance. Here we address their limitations for this purpose and identify the fishery data corresponding to the fish community off Newfoundland.

The use of fishery data in the study of ecosystems has, to some extent, been achieved in the Great Lakes (Keleher 1972; Regier and Loftus 1972; Regier 1973) as well as in some marine environments (Deimling and Liss 1994; Pauly et al. 1998). Recorded commercial landings are usually the only available data that go far enough back in time to cover the generation time of large species such as Atlantic cod. Landings may indicate the intensity of direct human disturbance on the community, although discards and bycatch are usually not reported. Fisheries catches usually indicate which are the most abundant species since those species are usually the most valuable from a commercial standpoint and are therefore targeted. The data can also show shifts in the relative abundance of species. Despite the fact that politicians, fishermen, economists and scientists disagree on the reliability of fisheries statistics, all at least agree that qualitative changes in target species are usually the consequence of a collapse in abundance of some fish stock (population).

Fisheries information is a reflection not only of the state of the stocks, but also of human decision-making. Fisheries effort intensifies when the abundance of species decreases in order to maintain the same level of catches over time, and fisheries technology enables the effective targeting of areas in which high levels of fish abundance may remain. As a result, fisheries data do not represent the real abundance of fish species. The index CPUE (catch per unit effort) homogenizes fishery data in order to compare the state of the fish species over time.

Table 2.1. List of demersal groups (plus pelagic species capelin and Arctic cod) that make up the commercial catches off Newfoundland (NAFO areas 2J 3K) during the years from 1960 to 1994

Species Name	NAFO Species Code
Atlantic cod	101
Haddock	102
Atlantic redfishes	103
Pollock	106
American plaice	112
Witch flounder	114
Yellowtail flounder	116
Greenland halibut	118
Atlantic halibut	120
Winter flounder	122
Flatfishes	129
American angler (goosefish)	132
Cusk	144
Greenland cod	148
Roundnose grenadier	168
Roughhead grenadier	169
White hake	186
Wolffishes	188
Groundfish	199
American eel (reported)	308
Baird's slickhead	326
Capelin	340
Dogfishes	459
Large sharks	469
Skates	479
Finfishes	499

However, the difficulty in calculating effort effectively produces a high error probability for the index value, which has discouraged its use in quantitative analysis of stock abundance (Hall 1999). In addition, fishery data sometimes may not permit resolution to the species level. This is the case in the Newfoundland 2J3K fishery data, in which some landing groups refer to a single species while some others can be very broad, for example reported simply as finfish or flatfish. However, the classification used here has been consistent through time, thus allowing comparison between years.

We have defined our fish community as those demersal species commonly occurring in the Newfoundland continental shelf ecosystem, plus those species that might not be demersal but are of primary importance as prey species. The classification groups according to common names are presented in Table 2.1. This table does not contain as separate classification groups eight demersal species that are present in the NAFO classification of landings: silver hake, red hake, blue ling (rare but reported), lumpfish, eelpouts, sculpins, argentines, and spiny dogfish. Since individuals of these eight species appear in less than three non-consecutive years in the catches, each was included in their correspondent indeterminate group: Groundfish or Finfishes. The pelagic species capelin and Arctic cod were included because of their trophic importance within the fish community.

Scientific Survey Data

Autumn surveys (mainly from October to December), conducted by Canada's Department of Fisheries and Oceans (DFO), collected the scientific data used in our study (Villagarcía 1995). Sampling was performed with an Engels trawl using a stratified random method with stratification by latitude, longitude, and ocean depth in a range of 100 to 1000 metres. Each stratum contained at least two sampling stations, each comprising a 30-minute tow at 2.5 knots with a codend mesh of twenty-nine millimetres. The survey was designed to provide information on species abundance. Samples were distributed evenly in time and randomly in space to allow standardized statistical analyses to determine the state of the fish community.

The way species were selected for inclusion in the analysis is explained below and the resulting fish community is shown in Table 2.2. From the data of the annual surveys, the number of individuals and total weight of each species were recorded. Data were standardized for duration of the tow and for number of stations surveyed each year. Thus, row data correspond to abundance as number or biomass per 30-minute tow and per year.

In order to eliminate sporadic or uncommon species whose presence will not significantly affect the analysis of interactions in the demersal community, species were excluded from the analysis if: (1) individuals were not identified to the species level, (2) they were pelagic (with the exception of capelin and Arctic cod), (3) they were present in only one year, or (4) if in half of the years in which the species occurred their abundance was less than five individuals, and their abundance in those years never comprised more than 0.05% of the total catch in that year.

As noted previously, despite the fact that capelin and Arctic cod are not demersal species they were included in the study due to their importance in the demersal trophic web. In scientific surveys, the total number of fish by tow is recorded and individuals are identified to the level of species whenever possible.

Table 2.2. List of the thirty-three species selected as components of the demersal community from scientific survey data off Newfoundland (NAFO areas 2J 3K) during the 1978 to 1993 period

Species Scientific Name	Canadian (English) Name
Agonus decagonus	Northern Alligatorfish
Anarhichas denticulatus	Broadhead Wolffish
Anarhichas lupus	Spotted Wolffish
Anarhichas minor	Striped Wolffish
Antimora rostrata	Blue Hake
Aspidophoroides monopterygius	Common Alligatorfish
Bathyraja spinicauda	Spinytail Skate
Boreogadus saida	Arctic Cod
Centroscyllium fabricii	Black Dogfish
Coryphaenoides rupestris	Roundnose Grenadier
Cottunculus microps	Arctic Deepsea Sculpin
Cyclopterus lumpus	Common Lumpfish
Eumicrotremus spinosus	Spiny Lumpfish
Gadus morhua	Atlantic Cod
Gadus ogac	Greenland Cod
Glyptocephalus cynoglossus	Witch Flounder
Hippoglossoides platessoides	American Plaice
Hippoglossus hippoglossus	Atlantic Halibut
Lycodes esmarki	Esmark´s Eelpout
Lycodes reticulatus	Arctic Eelpout
Lycodes vahlii	Vahl´s Eelpout
Macrourus berglax	Roughhead Grenadier
Mallotus villosus	Capelin
Myoxocephalus scorpius	Shorthorn Sculpin
Nezumia bairdii	Common Marlin Spike
Notacanthus chemnitzi	Large Scale Tapirfish
Amblyraja radiata	Thorny Skate
Malacoraja senta	Smooth Skate
Reinhardtius hippoglossoides	Greenland Halibut
Sebastes marinus	Golden Redfish
Sebastes mentella	Deep Water Redfish
Synaphobranchus kaupi	Longnose Eel
Trachyrhynchus murrayi	Roughnose Grenadier

Nonetheless, having different lifestyles, each species shows a different response to the sampling method. Capelin and Arctic cod are two species underrepresented in the data because they are pelagic species taken in a groundfish survey. That results in a certain bias in the representation of species in the sample that must be taken into account when discussing results.

Approaches in the Study of Fish Community Dynamics

There are many possible ways to study fish community dynamics. As Rice (2000) points out, single indices, multivariate analysis, descriptive curves and models can complement each other in obtaining a global picture when applied together to fish–fisheries interaction studies. Our goal was to show the changes in community structure over time and how these changes are affected by externalities. From that perspective we discuss next the usefulness of several common approaches.

Abundance and biomass are traditional measures that summarize the state of fish populations for fishery purposes. They are complemented with length at age, length frequency, fecundity at age, or sex ratio data. Despite their use to show changes over time in marine populations, these indices do not explain whether the observed changes accompanied changes in age structure or even if age structure changed regardless of a constant value in an index over time. Thus, abundance and biomass indices by themselves do not show the real state of populations. For example, a population can keep a constant abundance, despite the fact that it may be losing its oldest and largest individuals and decreasing in average size of individuals, as long as recruitment is enough to balance the loss of large individuals.

A combination of abundance and biomass indices to yield the average size of individuals in a population gives more information about its real state, and this information is especially relevant to fisheries. Changes of average weight (biomass/abundance) over time for a population and its relation to changes in abundance can indicate whether the adult individuals of the population are decreasing or increasing in number. This is important information when considering the capacity of the population to produce new individuals or to recover from a disturbance.

When applied to individuals instead of populations, size considerations also allow a community approach because size groups can be considered regardless of the species that individuals might belong to. In addition, as will be explained below, size can be used to build biomass and abundance spectra representing the community as a whole.

At the community level, there are other indices used to represent the state of the community. The most popular of those indices is species diversity. However, the use of diversity as an indicator of the community state is debatable, and controversy concerning theory and practice is extensive (Pimm and Hyman 1987). In an open environment, changes in diversity can result from the migration of species. That means that a change in abundance can be compensated by migration of new individuals into the community yielding no overall change in diversity despite a serious change in community structure and composition. Spatial density-dependent aggregation of individuals, which may mislead in the interpretation of diversity values, is also likely to occur in demersal environments. An additional drawback in the use of community diversity indices is that they can be misleading in the same way as happens with abundance and biomass indices. The same value of any of these indices can correspond to very different structural states.

The search for a holistic approach to the fish community has popularized the application of multivariate analyses, which operate by considering all the species of the community simultaneously. Martinez Murillo (2003) applied three different multivariate analyses to the demersal fish community data from 2J3K: cluster analysis, multidimensional scaling (MDS) and principal components analysis (PCA). Cluster analysis associates years or species that are similar in some respect. MDS not only associates, but also orders the variables. PCA allows identification of important factors that underlie the pattern observed over time. All these analyses reveal changes in the community structure. However, they are exploratory analyses, which means that they identify patterns but cannot confirm the reasons for them. Therefore, the outcomes from these methods should be complemented by confirmatory analysis (e.g. regression, rank correlation) to determine whether they might help to explain the changes observed.

Size spectrum analysis goes beyond the population to the community as a unit, by considering individual characteristics of the organisms. A size spectrum is the distribution of an attribute of a community as a function of size. Hence, individuals are grouped together according to the size stage they are in and regardless of the species to which they belong. The shape of the size spectrum and its change over time can give information about the ecological state of the community.

The more complex a system is, the more difficult it is to apply experimental tests to it. Indirect effects and time lag effects among variables are difficult to find using statistical models (Akenhead et al. 1982; Henderson 1987). Experimental tests assume an understanding of the system in order to be able to set the experiment and to give cause-and-effect results. Furthermore, from practical considerations, species with a long lifespan and a wide habitat range are difficult to deal with experimentally.

Many ecological phenomena are complex, include many variables, and are not completely understood. In these cases a model is most useful to put the pieces together and show the possible expected outcomes. Simulation models are appropriate for dynamic studies in which the processes themselves are equally as important as the final outcomes. Furthermore, a model can cover temporal scales that other approaches do not and can be very valuable when studying a large spatial scale where experimental replications are not possible (Carpenter et al. 1995). Models are very helpful in complementing experimental studies and identifying possibly important processes or factors that should be looked at more closely. In addition, a model allows the incorporation of new information about a system that may not have been considered or available otherwise. Although all these reasons support a simulation exercise, it is worth discussing to what extent models provide useful information.

Expectations from a Model

Most of the methods mentioned in the last section provide some abstract representation of the community; thus they are models themselves. Nonetheless, all are static and none of them can test the long-term dynamics of the fish community under different scenarios. A simulation model not only attains these objectives, but also permits integration of all levels of organization from individual to community and, without losing resolution at these levels, defines the emergent properties that come from their integration. The model we advocate (Martinez Murillo 2003) is size-based and deals with the interactions between individuals of the fish community. Because the abundance of offspring will influence the future abundance of large adult individuals, and at the same time the abundance of large adult individuals will influence that of offspring, a continuous cycle is established with no specific bottom-up or top-down structure that could influence the results.

Our simulation model for the Newfoundland demersal fish community (Martinez Murillo 2003) is intended to serve as a tool to: (1) study the importance of size in the structure of a fish community, and (2) quantify sensitivity to externalities in the fish community.

The difficulty of obtaining accurate quantitative results already mentioned stems from: (1) background "noise," as for example environmental uncertainty, which creates constant uncertainty (Beddington et al. 1984; Larkin 1996), (2) the absence of universal laws in ecology and the uniqueness of each ecological situation (Roughgarden 1998), (3) past scenarios that may be completely different from future ones, i.e. it will not work to model the asymptotic part of a sigmoidal curve using the equation that models the exponential growth of the same curve, and (4) data may be insufficient to give exact outcomes, e.g. many survey data do not cover the lifetime of species. Nonetheless, a model may still be useful even when it is not completely accurate. As Roughgarden (1998, ix) points out:

> Think of cooking. In most dishes the ingredients don't have to be measured to a milligram, nor the baking timed to the millisecond. A model too doesn't have to get everything exactly right, because it may still account for what is going on pretty well.

Trends, magnitude, and relative abundance are more robust measures than absolute values of the variables for a model result. Therefore, it may be better to look for significant changes in these measures rather than at exact abundance values when interpreting the results from a demersal fish community simulation model. The analysis of these measures can reveal: (1) emergent properties at the community level induced by the size factor, (2) indirect effects, i.e. the links between elements are as important as the elements themselves, (3) possible drastic change in the dynamics of the system, and (4) a framework for fisheries management, i.e. a general approach to consider fisheries resources and the effect of fisheries.

General modelling problems of balancing complexity with prediction error as well as generality with accuracy (Clark 1984; Sugihara 1984; Puccia and Levins 1985) are easier to deal with when the modelling process is viewed as an approach to understanding ecological phenomena. Again Roughgarden (1998, xi):

> Think of building a bridge at the mouth of a river. There's no universal bridge—one size doesn't fit all, but civil engineering offers a general approach to building bridges. Similarly, no two lakes are the same . . . Through modelling one can present the information about different systems in a common format, and see general features emerge.

The most important factors for the outcome of a model are the variables selected and assumptions considered in the construction of the model (Lai and Gallucci 1988; McAllister and Kirkwood 1998). The underlying assumptions in the simulation of the fish community under study here are that predation is the main interaction among individuals and that it is both density and size-dependent. These assumptions are well recognized in the literature

(Dunn 1979; Bax 1991, 1998; Yodzis and Innes 1992). In its basic form the model is deterministic, but it can incorporate internal or external stochasticity as shown by Martínez Murillo (2003) in her presentation of a series of likely scenarios for Newfoundland's demersal fish community.

To Sum Up

The fish community defined by its populations of interacting species sets the biological limits within which all production and any disturbance must take place. The practical importance for fisheries demands more than a description of the fish community; it requires an ability to quantitatively measure the standing stock and to quantify changes over time. Uncertainties associated with the available data limit the accuracy of strictly quantitative measures, and the reality of uncertainty is a compelling reason to look for overall trends instead of quantitative accuracy in fish–fisheries interactions. For this reason, many international organizations are adopting a precautionary approach that looks for limits and target reference points in fisheries management (NAFO 1997).

Multivariate methods and size spectrum analysis, reflecting changes occurring throughout the whole community, may indicate whether base points for future reference can be defined at the community level. Based on the structure and links among individuals within the fish community, a simulation model can also serve as a framework to observe the trends followed by the community under different management scenarios.

In this chapter we have presented a number of aspects that should be considered in approaching the study of dynamics in a fisheries ecosystem. Interacting fish species form a community, and food web interactions (largely governed by size) determine the overall community properties, including production. We suggest a holistic approach to this study of fish dynamics, argue for the importance of size as a factor determining functional groups in a community, and discuss how a size-based model can serve as a framework for the integration of biological knowledge of fish communities with decision-making about resource exploitation.

In the doctoral dissertation that was supported by *Coasts Under Stress*, Blanca Martinez Murillo (2003) developed a novel and innovative size-based dynamic model using ©Matlab software. Her model has only three simple assumptions: (1) fish pass through a series of age-determined size classes through their life history, (2) big fish eat little fish, and (3) predation cannot drive species to extinction. The model was parameterized using the Newfoundland data described in this chapter, and runs appear to simulate the history of the Newfoundland fishery very well.

The model is stable over runs of centuries, and from a stabilized state is used in the dissertation to explore several scenarios involving environmental and fishery disturbances. Preliminary findings are that: (1) recovery to stable conditions after a disturbance can take hundreds of years, (2) the system recovers to previous conditions after an environmental disturbance, (3) the system may return to a less productive state after a fishing disturbance, and (4) the state of recovery after a combination of environmental and fishing disturbances is unpredictable.

What does all this mean as regards food security? First, we must recognize that a key element in food security is relative stability and predictability of supply. For many of the important marine foods familiar to us in Canada (Turner et al., Chapter 1, this volume), this situation no longer applies. Atlantic cod are sorely depleted and the same is true of eulachon, *Thaleichthys pacifica* (Musick et al. 2000); Pacific salmon, *Oncorhynchus* spp. (Nehlsen et al. 1991); cusk, *Brosme brosme*; bocaccio, *Sebastes paucispinis*; porbeagle, *Lamna nasus*; winter skate, *Leucoraja ocellata;* and Atlantic redfishes, *Sebastes* spp. Even deep-sea fishes are not immune (Devine et al. 2006). Our ecological investigation suggests that ecosystem level consequences propagated through the food web make full recovery of any of these unpredictable and perhaps even impossible over the long term. For their own survival, human communities dependent on such resources must focus at once on broad and stringent measures for their conservation. Such measures will need to include attention to habitat conservation and to supporting these species during each phase of their lifecycles from their larval and juvenile phases to their full maturity, especially their spawning and reproductive phases, and the full range of habitats they require throughout their lives.

In: Resetting the Kitchen Table
Editors: C. C. Parrish et al., pp. 33-49

Chapter 3

AQUACULTURE AND NEARSHORE MARINE FOOD WEBS: IMPLICATIONS FOR SEAFOOD QUALITY AND THE ENVIRONMENT NORTH OF 50

Christopher C. Parrish[1], Louise Copeman[2], Geert Van Biesen[3] and Joseph Wroblewski[1]

[1]Ocean Sciences Centre, Memorial University of Newfoundland, Canada
[2]Biology, Memorial University of Newfoundland, Canada
[3]Environmental Science, Memorial University of Newfoundland, Canada

ABSTRACT

When faced with poor catches of wild fish, fish farming could provide food security to coastal communities. However, the economic and health benefits of cultured seafood have to be balanced against their environmental costs and health risks. While small-scale farms or land-based operations are much better for the environment, they may not be economical unless environmental costs are factored in. On the other hand, environmental effects resulting from intensive culture of carnivorous species can be reduced if large farms are placed in locations that are carefully chosen with respect to oceanographic currents, local resources, and natural migration routes. In this Chapter we discuss the nutritional value and safety of seafood in the context of aquaculture development and its environmental footprint, highlighting chemical analyses of samples we collected off the northern shores of Newfoundland and Vancouver Island and the southern shore of Labrador. Seafood collected between 50° and 53° N had high levels of healthful omega-3 fatty acids, and the bivalves had the additional benefit of containing phytosterols. Fish farms however, had an organic footprint extending more than a kilometre away from the centre of the farm, suggesting organisms in this radius may receive beneficial, detrimental, and/or neutral organic compounds derived from the farming activity. The effects of the farms are also felt much further afield in terms of pressure on feed-grade fish and commodities such as the omega-3 containing fish oils that are essential for optimal health of both the farmed fish and the humans consuming them.

INTRODUCTION

The culture of aquatic organisms could be a cornerstone for the provision of food security to coastal communities. When faced with poor catches of wild fish, fish farming should provide these communities with "access to sufficient, safe, nutritious, and culturally appropriate food to meet their dietary needs and food preferences" (Turner et al., Chapter 1, this volume). Low-intensity aquaculture has been providing food security in developing nations for hundreds of years; however, as with agriculture's "green revolution" of the last century there is a "blue revolution" occurring in aquaculture, and with it comes environmental and social impacts.

Aquaculture has become a large-scale commercial industry in many parts of the world. Climbing from 13 million tonnes of fish and shellfish produced in 1990 to 31 million tonnes in 1998 (85% of which is in developing countries), fish farming is poised to overtake cattle ranching as a food source by the end of this decade (Brown 2000). In Canada, commercial aquaculture has grown rapidly from a small cottage industry in the early 1980s, and fish farming now occurs in every province, almost exclusively in rural communities. Canada's contribution was 128,030 tonnes in 2000 (Canada 2005). Salmonids are the major product largely as a result of former government sponsored salmon and trout enhancement programs. In coastal Newfoundland and British Columbia, Atlantic salmon (*Salmo salar*) aquaculture has been practised since the 1980s. Atlantic salmon continues to be the major product of the aquaculture industry on both coasts, although bivalve aquaculture has been growing rapidly, and in Newfoundland, the decline of Atlantic cod (*Gadus morhua*) stocks, with a moratorium declared in 1992, encouraged research into the possibilities of cod farming. There are now considerable efforts towards commercialization of cod aquaculture.

The techniques that are currently used for salmon and cod aquaculture grew out of methods designed to optimize use of precious resources or attempts to reverse their depletion. In the 1980s cod farming started in Newfoundland as a means of fattening up under-sized cod caught by fishermen, and of using male capelin for which there was no market. Subsequently it was also shown that this approach was biologically viable in the colder waters of Gilbert Bay, Labrador (Wroblewski et al. 1998). Salmon farming grew out of the hatchery techniques used for enhancing wild salmon runs, and farming began in BC in the 1970s with small locally owned enterprises. However, the economy of scale has led to intensive aquaculture and the creation of vertically integrated companies on both coasts.

Coastal communities often welcome aquaculture development initially, but in the Broughton Archipelago and Clayoquot Sound where many of BC's salmon farms are now located, there is strong opposition to current operations and any further expansion. In Gilbert Bay a new approach is being proposed in which a Marine Protected Area has recently been established that permits aquaculture providing there is no impact on the Bay. Grow-out cod farming (Wroblewski et al. 1998) should have excellent potential since it is a low-technology approach and gear construction, handling skills, and the required equipment are already present in local communities (Wells 1999; AMEC 2002); however, issues of commercial viability of aquaculture in Labrador remain to be addressed.

In this chapter we discuss the nutritional value and safety of seafood in the context of aquaculture development and its environmental footprint, highlighting chemical analyses of samples we collected off the northern shores of Newfoundland and Vancouver Island and the southern shore of Labrador. We collected plants and animals in Gilbert Bay, NL (53°N, 56°W), in Bonne Bay, NL (50°N, 58°W) and in the Broughton Archipelago, BC (51°N, 127°W), which we extracted for the determination of essential nutrients and biomarkers. We also performed bacterial counts on some samples. In addition, in Gilbert Bay we set up a small cod enclosure in order to find lipid biomarkers that could be used for measuring dispersal of organic material from finfish farms.

SEAFOOD QUALITY AND SAFETY

Seafoods are consumed for their distinct flavour and taste as well as for their nutritional and potential health benefits. The quality of the seafood is related to its appearance, odour, flavour, texture, and nutritional value, while its safety can be compromised by contamination with microorganisms or toxic chemicals. However, safety and quality are not always directly correlated: low quality seafoods can be safe to consume, and *vice versa*.

Once harvested, seafood may be affected by the growth of harmful and spoilage bacteria. Bacterial counts indicate the general microbial quality of the seafood and surrounding water as well as hygiene during handling, processing, and storage. We performed counts on wild blue mussels (*Mytilus edulis*) and cultured Icelandic scallops (*Chlamys islandica*) collected in Gilbert Bay (Khan et al. 2004) and found the levels to be the same or lower than those in mussels from a commercial site in Newfoundland. Other microorganisms that can contaminate bivalve molluscs can cause paralytic, diarrhetic or amnesic shellfish poisoning. Consumption of bivalves is the principal way in which these toxins of algal origin pose a health hazard to humans. This is clearly an issue for both wild and cultured bivalves and, indirectly, for finfish farming as there is the possibility that nutrients released from fish farms could contribute to algal toxicity.

The fat or lipid content of the seafood is intimately associated with several aspects of seafood quality. Most of the lipid in seafood occurs in the form of fatty acids and most of these are combined in other molecules such as storage triglycerides and membrane phospholipids. In our seafood samples from Newfoundland we found that 67%–86% of the lipid consisted of fatty acids and that most of these formed part of larger compounds (Table 3.1). Only 3%–17% of the total lipids contained fatty acids in the "free" or uncombined form. In seafood from cold-water environments, most fatty acids occur as unsaturated fatty acids containing double bonds (–C=C–). In our seafood samples, 75%–83% of the fatty acids were unsaturated and of these most were polyunsaturated (containing more than one double bond). Polyunsaturated fatty acids accounted for 34%–57% of the total fatty acids. These fatty acids, which are obtained mainly from the diet, play a vital role in animal physiology. On the other hand, the oxidation of polyunsaturated fatty acids (PUFA) in seafood contributes to the loss of flavour and texture of the product.

The nutritional quality of seafoods depends on the environment from which they originate, season of harvest, and method of cooking. Omega-3 fatty acids, mainly eicosapentaenoic acid (EPA) and docosahexaenoic acid (DHA) are prevalent in seafoods.

Table 3.1. Lipid composition of seafood from Bonne Bay, Newfoundland

	Atlantic cod *Gadus morhua*	Capelin *Mallotus villosus*	Horse mussel *Modiolus modiolus*	Rainbow smelt *Osmerus mordax*	Winter flounder *Pseudopleuronectes americanus*	Brook trout *Salvelinus fontinalis*
Total Lipids (% wet wt)	0.4 ± 0.1	3.1 ± 1.1	1.2 ± 1.1	1.8 ± 0.5	0.3 ± 0.1	1.0 ± 0.0
Triglycerides (% total lipids)	3.5 ± 2.7	64.5 ± 10.0	37.0 ± 5.7	57.3 ± 18.2	8.1 ± 9.0	49.8 ± 4.7
Free fatty acids (% total lipids)	14.6 ± 2.4	4.8 ± 2.1	16.6 ± 1.6	2.8 ± 0.7	13.4 ± 4.3	6.1 ± 1.7
Phospholipids (% total lipids)	62.7 ± 5.8	22.2 ± 6.0	28.9 ± 5.0	29.8 ± 18.7	58.9 ± 8.1	31.1 ± 1.8
Total fatty acids (% total lipid)	67.9 ± 2.9	86.1 ± 2.9	77.9 ± 2.0	81.2 ± 4.4	67.0 ± 3.2	79.9 ± 2.8
Total unsaturated fatty acids (% total fatty acids)	76.7 ± 2.1	83.4 ± 0.8	74.6 ± 2.6	76.8 ± 1.9	74.2 ± 1.3	77.6 ± 3.7
Total PUFA (% total fatty acids)	56.2 ± 3.4	34.1 ± 5.2	48.6 ± 2.1	36.5 ± 2.4	57.4 ± 2.7	48.3 ± 5.1
Total ω3 fatty acids (% total fatty acids)	51.2 ± 3.2	30.6 ± 5.4	38.1 ± 1.5	32.4 ± 3.1	47.3 ± 3.3	43.0 ± 5.2
Total bacterial fatty acids (% total fatty acids)	1.8 ± 0.9	0.9 ± 0.16	4.3 ± 1.5	1.5 ± 0.6	2.7 ± 0.6	2.0 ± 0.4
Total zooplankton fatty acids (% total fatty acids)	3.1 ± 1.2	31.1 ± 4.7	2.8 ± 0.2	5.0 ± 1.7	1.9 ± 0.7	4.8 ± 0.5

In fact, most of the polyunsaturated fatty acids in animals from the marine environment are of the ω3 type. In our seafood samples from Newfoundland we found that ω3 polyunsaturated fatty acids accounted for 31%–51% of the total fatty acids, while in the colder waters of Gilbert Bay, Labrador levels were as high as 59% of total fatty acids (Copeman and Parrish 2004). These fatty acids have beneficial effects on plasma lipids and lipoproteins, cardiovascular disease, cancer, adipose tissue mass, and inflammatory diseases (Turner et al., Chapter 1, this volume), and the dose required can be quite modest. For example, consumption of 1.3 grams of EPA plus DHA per week significantly lowers the risk of cardiac arrest as a result of heart disease (Siscovick et al. 2000). This amount is in line with health Canada's dietary reference intakes (DRI) for adult men (www.hc-sc.gc.ca/hpfb-dgpsa/onpp-bppn/diet_ref_e.html). Table 3.2 shows the amount of seafood from our three study sites required to supply 1.3 grams of EPA plus DHA. The amount varied considerably according to species, anatomical location, and geographical location, and occasionally there were significant differences for a species according to year or site within a location. The liver values for "golden cod" (Gosse and Wroblewski 2004) were significantly different because some were from animals with seriously depleted energy reserves as they very likely had just started eating again after several months of starvation during winter. Cod store their fat in the liver and during winter this is used as a source of energy and for the development of the gonads. The significant differences among mussel samples from British Columbia and the fact that there were also significant differences with samples from Newfoundland and Labrador may relate to their proximity to fish farms in the Broughton Archipelago (see below).

Generally the amount of fish flesh required for the weekly dose of 1.3 grams ranged from about one hundred grams to two kilograms (Table 3.2) with less being required of fatty fish such as capelin (Table 3.1). If most of the bivalve samples from British Columbia and a horse mussel (*Modiolus modiolus*) sample are excluded, then the range for bivalves is about 200 grams to 1 kilogram. Thus depending on species and location, the lowering of the risk of cardiac arrest can be achieved with about one to twenty servings per week. Much less fish liver would be required, and particularly notable is rock cod (*Gadus ogac*) liver, which suggests this could form the basis of a nutriceutical industry.

In addition to being rich in healthful ω3 polyunsaturated fatty acids, shellfish such as mussels and lobsters contain about half the amount of cholesterol as found in the same weight of lean hamburger or steak (Krzynowek 1985). However, bivalve molluscs such as blue mussels contain an abundance of plant sterols or phytosterols derived from the algae they eat, as well as low levels of cholesterol (~50 mg per 100 g or 0.05%) compared to meat. Various nutritional and clinical studies have indicated that certain phytosterols can reduce plasma cholesterol concentrations thereby lowering the risk of cardiovascular disorders. Phytosterols inhibit the uptake of cholesterol and are themselves poorly absorbed (Hicks and Moreau 2001). We found that the major phytosterols, sitosterol, stigmasterol, and campesterol were present in surf clams (*Spisula solidissima*), Greenland cockles (*Serripes groenlandicus*), blue mussels (*Mytilus edulis*), and Icelandic scallops (*Chlamys islandica*) from Gilbert Bay, with these three sterols alone contributing 5%–20% of the total from up to twenty sterols (Copeman and Parrish 2004). Sitosterol, stigmasterol, and campesterol are also present in vegetables, cereals and nuts, but seafood has the added benefit of containing high levels of DHA and EPA.

**Table 3.2. Amount of seafood required to be consumed for a 1.3-gram dose
of EPA plus DHA**

Species	Location and Year	Wet weight (g) needed
Finfish muscle or liver		
Brook trout (*Salvelinus fontinalis*)	Newfoundland 2003	517 ± 88
Capelin (*Mallotus villosus*)	Labrador 2000	109
	Newfoundland 2003	202 ± 60
Golden cod (*Gadus morhua*) flesh	Labrador 2000–2002	588–625
Golden cod (*Gadus morhua*) liver	Labrador 2000–2002	20–39*
Herring (*Clupea harengus*) flesh	Labrador 2000	134 ± 5
Herring (*Clupea harengus*) liver	Labrador 2000	129 ± 4
Northern cod (*Gadus morhua*) flesh	Labrador 2000–2001	563–838
	Newfoundland 2003	880 ± 487
Northern cod (*Gadus morhua*) liver	Labrador 2000–2002	25–31
Rock cod (*Gadus ogac*) flesh	Labrador 2000	544 ± 20
	Newfoundland 2003	598 ± 177
Rock cod (*Gadus ogac*) liver	Labrador 2000	17 ± 0.5
Smelt (*Osmerus mordax*)	Newfoundland 2003	334 ± 81
Winter flounder (*P. americanus*)	Newfoundland 2003	1776 ± 653
Bivalve organism or adductor muscle		
Blue mussel (*Mytilus edulis*)	Labrador 2000	944 ± 54
	Newfoundland 2003	443 ± 348
	British Columbia 2003	858–8319*
Horse mussel (*Modiolus modiolus*)	Newfoundland 2003	4616 ± 4401
Greenland cockle (*Serripes groenlandicus*)	Labrador 2000	1108 ± 28
Icelandic scallop (*Chlamys islandica*)	Labrador 2000	821 ± 39
Surf clam (*Spisula solidissima*)	Labrador 2000	728

*Significant differences among sites or years

No matter how good the seafood is for you, it has no nutritive value if it is not eaten. In countries where choice of food is abundant, palatability becomes a major factor in determining the food that is eaten and therefore nutrient intakes. Fats contribute to palatability in two ways, firstly by responses to their texture or "mouthfeel," and secondly by the taste in the mouth and aroma in the nose.

Lipids are also the dominant factor in determining organic contaminant accumulation in aquatic organisms. Contaminants move though the food web when contaminated prey are ingested. Trophic transfer can lead to biomagnification when contaminant loads increase significantly with each successive trophic level. This occurs with lipophilic ("fat-loving") contaminants that are poorly metabolized, such as polychlorinated biphenyls (PCBs). Ingestion of contaminated food (*cf.* Table 3.3) is also the principal way in which humans are exposed to these highly toxic environmental pollutants that are distributed worldwide.

A recent paper in *Science* on PCBs in farmed salmon (Hites et al. 2004) has generated considerable media attention to risks associated with consumption of Atlantic salmon, even though levels are lower than the Food and Drug Administration's tolerance levels of 2000 parts per billion (ppb) or nanograms per gram (ng/g) (www.cfsan.fda.gov). Attempts have been made to discredit the study on the basis of the researchers having used skin-on fillets, although essentially similar results were obtained with a smaller sample of skinned fillets earlier (Jacobs et al. 2002).

Table 3.3. Polychlorinated biphenyl and polyunsaturated fatty acid concentrations in salmon and some other common foods

	PCB			PUFA
	ng/g lipid[1]	ng/g food[1]	ng/serving[2]	g/serving[2]
Salmon[3]	300 (145–460)	32.9	2790	2.0
Canned tuna[4]	3.9	0.2	13.3	1.2
Chicken[4]	46 (41–52)	3.1	170	0.9
Chicken liver[4]	68	4.4	88	0.2
Butter[5]	1.7 (0.5–3.8)	1.4	20	0.4

[1]The concentration unit is the same as ppb. To convert from data normalized to lipid, to data normalized to g food, USDA data (mostly averaged) were used (www.nal.usda.gov/fnic/foodcomp/Data/SR16-1/wtrank/wt_rank.html).

[2]Values per serving were calculated from USDA data (www.nal.usda.gov/fnic/foodcomp/Data/SR16-1/wtrank/wt_rank.html). Portion sizes were comparable to those of the Canadian Food Inspection Agency (www.inspection.gc.ca/english/bureau/labeti/guide/5-0-1e.shtml).

[3]Salmon PCB data are an average (range in brackets) of 8 wild and farmed Atlantic salmon from Europe (Jacobs et al. 2002). Most of the PUFA are the long-chain ($\geq C_{20}$) $\omega 3$ fatty acids EPA and DHA (Ackman, 1989; www.nal.usda.gov/fnic/etext/macronut.html).

[4]Tuna, chicken, and chicken liver PCB data are from an industrial part of Spain (Bordajandi et al. 2004) and the chicken data are an average (range in brackets) of 2 samples.

[5]Butter PCB data are an average (range in brackets) of 6 samples from Canada (Kalantzi et al. 2001). Most of the PUFA are short-chain (C_{18}: www.nal.usda.gov/fnic/etext/macronut.html)

In addition Hites *et al.* have been criticized for not putting their data into the context of human food consumption patterns, and the *Seattle Post–Intelligencer* (15 January 2004) generated a chart showing butter, tuna, and chicken to have higher levels of PCBs than some farmed salmon. This chart has been used by aquaculture promoters, including academics, but a survey of some of the primary scientific literature (Table 3.3) shows that opposing data also exist. This also points out the problems associated with this kind of information in that different units and conversion factors are applied, confounding the interpretation.

Since PCBs are fat soluble, the analytical data are usually reported as nanogram per gram lipid, although Hites *et al.* reported their data per gram wet weight. Conversion factors can then be used to calculate PCB concentrations per gram of food and per serving (Table 3.3) as well as consumption advisories (Hites et al. 2004). Nonetheless, the PCB concentration data for salmon (and ooligan grease: Chan et al. 1996) remain below the US FDA's tolerance level and, as aquaculture promoters point out, the amount of salmon eaten per capita is comparatively small. Thus, while salmon have higher PCB concentrations than other foods (Table 3.3) including beef and milk, on average we obtain more PCBs from the latter because we consume more poultry, beef and milk in a year. This low salmon consumption seems to be an ironic way of promoting salmon farming, as is the amount of long-chain $\omega 3$ fatty acids when proportions are lower than in wild salmon (Olsen 1999; *www.nal.usda.gov/ fnic/etext/macronut.html*) and industry is trying to reduce them further. These fatty acids are supplied by fish oils, which are one of the most expensive components of fish feed, and demand for this commodity is rapidly outpacing fish oil supply.

NEARSHORE MARINE FOOD WEBS

The nearshore environment provides a very diverse range of habitats for living organisms, but they are frequently at risk of environmental degradation due to proximity to anthropogenic influences. These habitats can act as critical biological transition zones (Levin et al. 2001), which are broadly defined as "the interfaces between soils, fresh water sediments, and marine sediments that can control or influence the movement of organisms, nutrients, energy, and pollutants within and across landscapes" (SCOPE 1999). We have been investigating local food web linkages in such nearshore habitats off the coasts of British Columbia and Labrador and the west coast of Newfoundland, in order to understand basic ecosystem functions such as the recycling of terrestrial nutrients and organic matter and the use of these nutrients in benthic communities.

In order to investigate these biological processes, we have used a chemical approach that involves the analysis of lipid bio-indicators. Useful lipid biomarkers are those compounds that are formed at the base of the food chain and are transferred relatively unchanged throughout higher trophic levels (Reuss and Poulsen 2002). This approach has previously been used to describe the transfer of primary production throughout the food chain (Fraser and Sargent 1989), to assess the suitability of a marine site for aquaculture development (Budge et al. 2000), or to assess the impact of aquaculture on the marine environment (Henderson et al. 1997). In the context of aquaculture development in coastal Canada, we have been investigating two study sites, Gilbert Bay and the Broughton Archipelago. Gilbert Bay is a shallow brackish bay, located along the Labrador coast (52° 35'N 56° 00'W) near the communities of William's Harbor and Port Hope Simpson. The Broughton Archipelago is a group of islands located off the northeast coast of Vancouver Island, north of Johnstone Strait (50° 43'N 126° 34'W).

Nearshore benthic communities efficiently process much of the nutrients from both terrestrial and planktonic sources. In Gilbert Bay, we found that sixteen species of macroinvertebrates had elevated proportions of polyunsaturated fatty acids (PUFA: 48.0 ± 7.3%), which varied surprisingly little across the species, suggesting that they were not dependent on feeding modes. This finding emphasizes the importance of PUFA to physiological function in ectotherms living in cold water: the PUFA help biological membranes to remain fluid at cold temperatures. Further, levels of PUFA found in macroinvertebrates in both the Broughton Archipelago and Gilbert Bay are higher than those found in surrounding sediments (Copeman and Parrish 2003) and equally as high as those found in plankton (Table 3.4). This demonstrates that these invertebrates are able to concentrate and retain essential lipids from their diet. Invertebrates also perform vital functions related to processing lipids within sediments. In Gilbert Bay, levels of PUFA were found to be elevated in the freshly deposited sediments termed "fluff" (Copeman and Parrish 2003) compared to those found in the older sediment layers. Previous studies have demonstrated that the presence of macroinvertebrates in sediments significantly changed the vertical distribution and degradation rates of lipids (Sun et al. 1999).

Table 3.4. Lipid biomarkers in plankton and bivalves from various locations in Gilbert Bay, Labrador, and the Broughton Archipelago, British Columbia

	Gilbert Bay Plankton	Broughton Plankton	Gilbert Bay Blue mussels *Mytilus edulis*	Broughton Blue mussels *Mytilus edulis*
Total Lipids (% by weight)	4.3 ± 1.6 (% dry wt)	6.0 ± 3.1 (% dry wt)	0.6 ± 0.1 (% wet wt)	0.4 ± 0.3 (% wet wt)
Total PUFA (% total fatty acids)	50 ± 4.7	30 ± 14	57 ± 2.2	51 ± 9.3
Total ω3 fatty acids (% total fatty acids)	39 ± 5.8	22 ± 12	44 ± 1.2	36 ± 8.8
Total bacterial fatty acids (% total fatty acids)	5.0 ± 2.3	21 ± 10	2.8 ± 0.2	7.4 ± 2.9

ANTHROPOGENIC THREATS TO COMMUNITIES

The worldwide growth in aquaculture production has increased awareness of its environmental impacts. This has already led to moratoria on new developments and tighter control in several countries (Wu 1995). The potential environmental impacts of aquaculture cover a wide array of topics. Depletion of fish stocks used as fish feed (Naylor et al. 2000); the destruction of coastal ecosystems which are often important nursery grounds for marine species (Iwama 1991); eutrophication caused by nutrients emitted directly from the fish and from the decomposition of excess food and faeces (Aure and Stigebrandt 1990); anoxic conditions in the sediment caused by organic enrichment and affecting benthic communities (Tsutsumi et al. 1991), the use of antibiotics (Bjorklund et al. 1989) and antifoulants (Davies and McKie 1987), and sea lice infestations (Morton et al. 2004; Volpe, Chapter 5, this volume) are some of the many possible concerns.

The most obvious concerns over impacts of intensive culture of carnivorous species are those on the local environment. The effects of the sedimentation of food particles and faecal pellets on the plants and animals on the seafloor may be significant. A high input of organic detritus to the sediment can lead to anoxic conditions, which in turn can lead to the production of toxic gases such as ammonia, hydrogen sulfide, and methane (Hall et al. 1990; Wu 1995; Ervik et al. 1997). These chemical changes often go hand in hand with a shift in the benthic community (Tsutsumi et al. 1991; Hansen et al. 2001). Typically, under undisturbed conditions the competitively dominant species are those large-bodied, long-lived species with a relatively stable population size close to the carrying capacity of the environment. Smaller and shorter life-span species may be numerically dominant, but they contribute a relatively small proportion of the biomass. If severe or frequent disturbances occur, the conservative large-bodied, long-lived species are the first to succumb while the opportunistic small body size and short life-span species are favoured and diversity decreases. In time, the opportunists may become dominant in terms of both biomass and numbers (Ritz et al. 1989). Any of these chemical or biological effects could be compounded by local hydrography, cold temperatures or the sensitivity of the nearshore habitat.

Gilbert Bay is a sheltered body of water that is ice covered for six months of the year and has constant benthic temperatures (at >50 m) of −1.5°C (Green and Wroblewski 2000; Morris et al. 2002). It has recently been designated as a Protected Area in the Marine Protected Areas Program (DFO, Canada) primarily because of the genetically distinct Atlantic cod (*Gadus morhua*) population that inhabits the bay (Ruzzante et al. 2000; Morris et al. 2002). However, Gilbert Bay has long been used as a traditional fishing ground by the Métis people and has also been a pilot site for the development of Atlantic cod (*Gadus morhua*) and Icelandic scallop (*Chlamys islandica*) aquaculture industries (Green and Wroblewski 2000). For carnivorous species, both decreased microbial activity at lower temperatures (Pomeroy et al. 1991) and low current flow could exacerbate the build up of organic material surrounding fish cages (Silvert 1992). Furthermore, it is likely that most of the organic matter from fish farms would reach the benthos in Gilbert Bay, compared to deeper sites, due to a shorter residency time within the water column.

Since 2002, the Broughton area in BC has received heightened media attention due to the collapse of pink salmon (*Oncorhynchus gorbuscha*) stocks as a result of sea lice infestation from the surrounding Atlantic salmon farms (Morton et al. 2004). However, aquaculture is likely also impacting juvenile fish and bivalves through indirect effects such as the destruction of critical nearshore habitat. Much of the nearshore habitat within the Broughton is characterized by kelp and eelgrass habitat. Eelgrass (*Zostera marina*) is a colonial marine flowering plant that occurs in shallow soft sediments. It is widely distributed globally in coastal waters and it is among some of the most productive marine habitat, forming vital nursery areas for both juvenile fish and invertebrates (Duarte 1989). However, eelgrass is currently under threat on a global scale due to nearshore anthropogenic activities such as aquaculture development (Short and Wyllie-Echeverria 1996). The overloading of nutrients from fish farms can indirectly cause degradation of eelgrass beds. This habitat is highly productive not only due to eelgrass but also due to epiphytic populations, sessile plants and animals that grow attached to their eelgrass hosts (Penhale and Thayer 1980). Epiphytes are often responsible for up to 30% of the total production of eelgrass beds, however, with increased nutrification it has been shown that epiphytes overgrow eelgrass resulting in reduced meadow productivity (Orth and Moore 1983). This is a consequence of shading of eelgrass blades, which decreases photosynthesis. Further, overgrowth by epiphytes can also limit nutrients available to the eelgrass plants. Through this mechanism, aquaculture poses further risk to nearshore habitat in the Broughton, indirectly threatening juvenile fish and bivalves that inhabit this highly productive environment.

Changes in food web structure due to aquaculture occur against a background of other impacts. In fact, anthropogenic changes on community diversity can now be seen in all marine systems on a global scale (Vitousek et al. 1997). The ability to predict human impacts on marine community diversity, and in particular, species abundance is complicated by the numerous indirect and direct interactions between predators and prey (Yodzis 1998). These impacts have previously been discussed in terms of a simplified method, which considered control of aquatic communities either through a 'top-down' or 'bottom-up' approach (McQueen et al. 1989). In the bottom-up theory, organisms on each trophic level are resource-limited and their population is thought to depend mainly on increased primary production. Therefore, in bottom-up theory increased eutrophication dramatically affects community structure and function. In the top-down approach, control of prey dynamics is by predators with only the top predator experiencing resource limitation. Predators then regulate the abundance of their prey at each successive trophic level (Estes et al. 1998). In top-down control the impact of predator removal is transmitted down the food web influencing lower trophic levels and ultimately regulating primary production through a trophic cascade. Although these theories provide a framework in which to organize complex observations of anthropogenic food-web alteration, in reality, marine communities are often influenced by both bottom-up and top-down impacts with unpredictable results to marine community diversity.

Human fishing pressure has led to a decrease in mean trophic level of fish catches on a global scale. Pauly et al. (1998) termed this reduction in trophic level as "fishing down the food web." Declining stocks of large piscivorous fish are thought to result from a combination of higher fishing effort and life history characteristics that make these species relatively more vulnerable. Jennings et al. (1998) looked at trends in abundance in eighteen exploited fish species from the northeastern Atlantic and found that after controlling for fishing effort and phylogeny, fish with later age at maturity and larger size were more likely to decrease in abundance. Thus, fisheries that target both larger piscivorous fish and smaller planktivorous species will lead to a greater reduction in higher trophic level species over time.

Pauly et al. (2001) recently examined these global trends in relation to Canadian aquatic food webs and found that Canadian waters had shown a dramatic decrease in trophic levels. There was a particularly dramatic decline on the east coast of Canada from 1950–96 reflecting the change in fisheries dominated by gadoids to domination by pelagic fisheries to domination by low-trophic level invertebrates such as shrimp and crab. Pauly et al. (2001) conclude that declining trophic level with declines in catches could infer a food web collapse. Further, it was concluded that biological production that was originally used in the fish food web could now be diverted towards alternate components of the ecosystem like small benthic invertebrates and jellyfish. One such example of this increase in invertebrate production with release from top-down control is the increase in northern shrimp (*Pandalus borealis*) stocks with the concomitant collapse of northern cod (*Gadus morhua*) during the last twenty years. Worm and Myers (2003) demonstrated through meta-analysis of nine fisheries regions in the Northwest Atlantic that shrimp biomass was strongly negatively correlated with cod biomass but was not correlated with temperature.

Eutrophication has wide ranging effects on marine community diversity and structure. However, anthropogenic eutrophication of marine habitat commonly leads to a decrease in species diversity with an increase in a few tolerant species (Smith et al. 2000). Therefore, in coastal areas of Canada with intensive aquaculture, food web alternations are occurring both due to top-down effects of overfishing but also due to bottom-up effects of eutrophication. This situation is one that leaves marine coastal areas vulnerable to colonization of invasive species such as gelatinous zooplankton. The Black Sea is a good example of this food web collapse coupled with intense blooms of the common moon jellyfish, *Aurelia aurita* and the Ctenophore *Mnemiopsis* (Daskalov 2002).

If regime shifts are occurring in coastal marine systems, what could this mean in relation to the human food chain? PUFA in seafood originates in primary production from photosynthetic algae. However, the biochemical form in which this PUFA is present is changed and concentrated as it moves up the food chain (Table 3.5). Declining nearshore species, such as higher trophic level white fish and bivalves, generally have very high proportions of PUFA (55%) and lean flesh (cod and scallops have 0.5 g lipid per 100 g wet weight). In particular, fish have elevated DHA while bivalves generally have elevated EPA.

These fish have low levels of cholesterol with bivalves having elevated sterols but also having a significantly increased level of phytosterols (Copeman and Parrish 2004). Lower trophic level pelagic fish have decreased proportions of PUFA (20%) in their flesh and higher overall levels of lipid (8 g per 100 g). Further, invertebrates such as shrimp and crab have elevated levels of PUFA but they also contain much higher levels of cholesterol than found in either white fish or shellfish. Little is known about the lipid content of jellyfish, but low levels of lipid have been found in their flesh and unique very long chain fatty acids can comprise over 12% of their total fatty acids (Table 3.5). Nothing is currently known about the effects of these fatty acids on human health. Therefore, from a lipid perspective, not all seafood is created equal and regime change could have an effect on both the quantity and quality of PUFA and phytosterols available for human consumption.

Table 3.5. Lipid composition of seafood from different trophic levels

	Jellyfish	Mussels	Crab	Shrimp	Herring	Capelin	Cod
Total Lipid/wet wt (g/100 g)	Very low	0.6	0.3	2	6.8	9.9	0.5
% PUFA	48[a]	58[b]	50[b]	?	20[d]	20[d]	58[d]
% DHA	10[a]	12[b]	16[b]	7[c]	9[d]	4[d]	33[d]
% EPA	14[a]	20[b]	22[b]	12[c]	7[d]	9[d]	19[d]
% Very long chain PUFA	12[a]	–	–	–	–	–	–
Cholesterol/wet wt (mg/100 g)	?	60[e]	70–120[e,f]	100–212[e,f]	?	?	35[f]
% Phytosterols	?	high	low	low	low	low	low

[a]Nichols *et al.* 2003, [b]Copeman and Parrish 2003, [c]Ackman 1989, [d]Copeman and Parrish 2004, [e]Krzynowek 1985, [f]Ackman and McLeod, 1988

Given that PUFA is in a limited supply, does the low-fat high-PUFA aquaculture product (cod and flatfish) justify the net loss of high quality lipids (capelin and herring) from the human food chain? These aquaculture products are particularly attractive to the health conscious public who want low fat products that still contain high levels of omega-3 PUFA. Pelagic species contain lower percentages of PUFA (20% of total fatty acids) compared to higher lean fish (50%) but the overall level of lipid is much higher in pelagic species (8 g lipid per 100 g wet weight). Thus, consuming lower trophic level fish would generally result in higher PUFA per unit flesh than higher trophic level aquaculture species. This fact is troubling from an environmental perspective, considering that the aquaculture industry is currently still dependent on feeding lower trophic level species (e.g. herring) to higher trophic level species (e.g. cod). Currently, higher trophic level aquaculture products still represent a net loss of edible PUFA in the human food chain.

THE ORGANIC FOOTPRINT OF FINFISH FARMS

Our studies of the impact of aquaculture activities focussed on the tracking of organic waste from fish farms in Gilbert Bay, Labrador and off the coast of Vancouver Island. These effects are not only important from an environmental point of view, but have direct relevance to the fish farm itself since deterioration of water and sediment quality can negatively influence its production. It is therefore important to know the extent of the area that is affected by the output of the aquaculture facility, or its footprint. The impacts of fish farming are highly dependent on local conditions, which means that for a given environmental loading, the impact can be quite different for different systems. A higher current velocity allows for settling material to be dispersed to a greater extent. Similarly, a greater depth under the cages allows prevailing currents to dilute and disperse the settling material to a greater extent than in shallower depths (Iwama 1991). Common sense dictates that fish cages should be put in well-flushed and not too shallow locations, but this can conflict with the requirements of the species being cultivated. Also, sheltered bays provide the cages with better protection from heavy weather.

The magnitude and nature of particulate matter generated by a fish farm, or the environmental loading, depends on a number of variables, such as the biomass of the farm, the species being cultivated and the kind of feed used (fish of low commercial value or pellets). The size of the individual organisms is also important since this determines feed and faeces particle size. Small particles (high surface to volume ratio) with a high moisture content will have the highest dispersion rate. Automatic feeding methods tend to waste more feed than feeding by hand, because with the latter method the rate at which the feed is administered can be adjusted to the rate at which the fish are feeding, and this can vary on a daily basis. Thus, it is not surprising that the literature reports a wide range for the amount of feed that is uncaptured: 1%–8% (Wu 1995; Stewart 1997). The values for the amount of faecal waste are more consistent, with a range of 20%–30% relative to the amount of feed consumed (Iwama 1991; Stewart 1997).

Typically, feed for salmonid culture is composed of ~50% protein, ~20% carbohydrate, ~15% lipid, and varying amounts of water and minor components such as antioxidants, vitamins, pigments, and therapeutic agents (Iwama 1991). Although the fate of individual proteins, carbohydrates, and lipids has not been studied extensively, Johnsen *et al.* (1993) and Henderson *et al.* (1997) analyzed lipids in the sediments underlying salmon cages in Scotland and Norway respectively, and were able to relate them to the feed. For instance, the principal fatty acid in the fish feed was present in the sediments underneath the cages and its proportion decreased along a transect away from the cages. Henderson *et al.* (1997) observed the same trend for total lipid and individual lipid classes.

Herring (*Clupea harengus*) or herring-based feeds are commonly used in aquaculture as a source of ω3 fatty acids for fish. In our 2001 enclosure study in Gilbert Bay we fed 128 kg (281 lbs) of herring, 33 kg (73 lbs) of capelin (*Mallotus villosus*), and 8 kg (18 lbs) of fish pellets (Haddock Grower 46–14, 6.5 mm, Zeigler Bros, Inc, Gardners, PA, USA) to 238 cod (Table 3.6). At first, the feeding of ω3 fatty acids to fish may seem paradoxical but, as noted previously, fish are themselves unable to synthesize ω3 fatty acids. The fish have to acquire them from preceding trophic levels within the food web, ultimately starting with microalgae or phytoplankton. Phytoplankton make fatty acids with certain ω designations (ω3, ω6 etc.) which cannot be modified by animals, yet some of these ω3 and ω6 fatty acids are essential to animal growth.

Table 3.6. Fatty acid composition (% total) of cod feed. Data are major fatty acids present at >1% in at least one of the sample types and are expressed as mean ± standard deviation for 3 samples except for the capelin which was a single sample caught just outside Gilbert Bay in August 2000

	Herring (*Clupea harengus*)	Haddock Grower (Zeigler Bros., Inc.)	Capelin (*Mallotus vilosus*)
14:0	4.3 ± 0.2	5.0 ± 0.1	5.9
16:0	11.7 ± 3.4	17.8 ± 0.3	8.7
18:0	1.1 ± 0.4	4.4 ± 0.1	0.6
Total saturated fatty acids	17.9 ± 3.7	29.4 ± 0.4	17.1
16:1ω7	8.1 ± 0.9	7.8 ± 0.0	10.3
18:1ω9	8.7 ± 5.1	16.6 ± 0.2	3.8
18:1ω7	2.4 ± 0.4	2.8 ± 0.1	1.6
20:1ω9	15.3 ± 6.0	5.8 ± 0.1	16.6
20:1ω7	1.3 ± 0.6	0.5 ± 0.0	1.0
22:1ω11	23.1 ± 9.9	7.6 ± 0.1	9.3
22:1ω9	1.4 ± 0.8	0.9 ± 0.0	18.5
Total monounsaturated fatty acids	61.1 ± 10.1	42.5 ± 0.5	62.6
16:4ω1*	0.6 ± 0.1	0.6 ± 0.0	2.0
18:2ω6	0.7 ± 0.3	9.3 ± 0.1	0.5
18:3ω3	0.3 ± 0.1	1.1 ± 0.0	0.2
18:4ω3	0.8 ± 0.2	1.3 ± 0.0	1.2
20:5ω3 (EPA)	7.2 ± 2.5	5.8 ± 0.1	9.3
22:6ω3 (DHA)	8.2 ± 3.2	5.8 ± 0.2	4.1
Total PUFA	20.9 ± 6.5	28.1 ± 0.6	20.0
Bacterial fatty acids*	0.8 ± 0.2	1.5 ± 0.2	1.8
Terrestrial plant fatty acids*	1.0 ± 0.4	10.3 ± 0.1	0.7
Total ω3 fatty acids	17.8 ± 6.2	15.8 ± 0.4	15.2
Zooplankton fatty acids*	39.8 ± 11.6	14.3 ± 0.1	44.4

*Biomarkers (Parrish et al. 2000):

Bacterial:	15:0 + 15:1 + *i*15:0 + *ai*15:0 + *i*16:0 + *ai*16:0 + 17:0 + 17:1 + *i*17:0 + *ai*17:0
Terrestrial:	18:2ω6 + 18:3ω3 > 2.5% means significant terrestrial input
Diatom:	16:4ω1
Zooplankton:	20:1ω9 + 22:1ω11 + 22:1ω9

In addition to their nutritional importance, these fatty acids can be used as food web tracers delineating the pathway between phytoplankton and fish. Zooplankton make some other long carbon chain fatty acids with designations such as ω9 and ω11. These are transferred to zooplankton-eating fish such as herring and capelin (Table 3.6). The large proportion of zooplankton markers in the commercial food suggests a significant contribution of fishmeal and/or oil from zooplanktivores; however, proportions were lower than in unprocessed capelin or herring. By contrast, fatty acids from terrestrial plants were much higher in the pellets reflecting inclusion of wheat, soy, and corn among the ingredients.

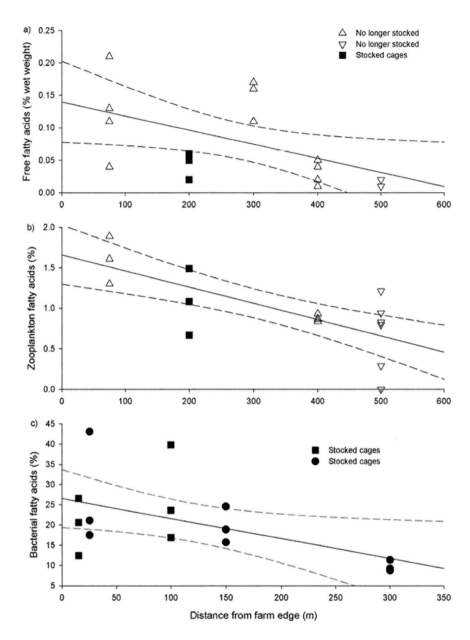

Figure 3.1. Fish farm lipid markers in marine animal and plant samples from the Broughton Archipelago, British Columbia. Different symbols represent different current and former Atlantic salmon farms. a) Blue mussels (*Mytilus edulis*). Data are free fatty acids as a percentage of total wet weight of individual mussels collected in the vicinity of current and former farms. b) Limpets (*Notoacmacea scutum*). Data are total zooplankton fatty acids as a percentage of total fatty acids in individual limpets collected in the vicinity of current and former farms. c) Plankton. Data are total bacterial fatty acids as a percentage of total fatty acids in repeat plankton tows conducted parallel to cages stocked with salmon

When cod or salmon consume the herring, they preferentially retain the ω3 fatty acids over the less important and less easily digested zooplankton fatty acids. These then become useful biomarkers of the dispersal of organic waste from fish farms. The form in which these molecules appear in the environment permits us to distinguish between uneaten food and faecal material. If they appear in the free form rather than combined into other larger molecules (e.g. triglycerides) then this represents faecal material. This was the case in our small cod enclosure study in Gilbert Bay, Labrador. Free fatty acid levels and the proportions of the long-chain monounsaturated fatty acids 20:1ω9, 22:1ω9, and 22:1ω11 in settling particles close to the pens (5 m) were significantly higher than before the fish enclosure was in operation, and than further away from the pens (Van Biesen and Parrish 2005).

The same organic compounds that we established to be useful biomarkers for tracking organic waste from the cod enclosure in Gilbert Bay were investigated off the coast of Vancouver Island. We sampled around current and former Atlantic salmon aquaculture sites in the Broughton Archipelago and found that the lipid fish farm markers together with bacterial markers were significantly correlated with distance from the edge of the farm site (Figure 3.1). The free fatty acid concentration in mussels decreased significantly with distance from the edge of the farm (Figure 3.1a). Overall, the average proportions of free fatty acids and zooplankton fatty acids were significantly higher in BC mussels than in Labrador mussels (Copeman and Parrish 2004) suggesting bioaccumulation from large salmon farms. This footprint extended over a kilometre from the farm centre. In fact, zooplankton marker proportions in mussels were at least two-thirds higher than in all other mollusc, plankton, seaweed, and sediment samples collected in the Broughton. Other molluscs may nonetheless still accumulate lipid fish farm markers albeit at lower levels. This was the case for proportions of zooplankton fatty acids in limpets which had lower proportions than in mussels but these proportions were significantly correlated with distance from the farm (Figure 3.1b).

In addition to the compounds we established to be useful fish farm markers in Gilbert Bay, we found that bacterial fatty acid levels were high in bivalve and plankton samples from the Broughton Archipelago, suggesting these data could also be useful for determining the footprint of fish farms. Overall, bacterial fatty acid proportions were significantly greater in mussels from the Broughton Archipelago than those collected in Labrador (Table 3.4) or Bonne Bay (Table 3.1). They were also very high in the water column within 150 metres of the farm edge (Figure 3.1c), and were at all times significantly higher than the highest level measured in Labrador (Copeman and Parrish 2003).

CONCLUSION

Clean water and the promise of jobs will undoubtedly continue to fuel aquaculture development north of 50°N. In the event of any significant increases in water temperatures, aquaculture in this region would become even more attractive because of increased growth rates, especially in the face of decreased aquaculture in Asia because of thermal stress. However, the economic and health benefits of cultured seafood have to be balanced against their environmental costs and health risks. While small-scale farms, as we looked at in Labrador, or land based operations are much better for the environment, they may not be economically viable unless environmental costs are factored in. On the other hand, environmental effects resulting from intensive culture of carnivorous species can be reduced if large farms are placed in locations that are carefully chosen with respect to oceanographic currents, local resources, and natural migration routes. There must also be adequate provision for fallowing and for monitoring of active and fallowed sites. In fact, mandatory measures to prevent environmental problems, such as requirements for fallowing, and improvement in monitoring are among many management and regulatory issues surrounding salmon aquaculture currently being considered by the BC government.

ACKNOWLEDGEMENTS

We are greatly indebted to Karen Gosse and Tracey Watson without whom the collection of the samples in Gilbert Bay and Bonne Bay would not have been possible. We would also like to express our gratitude to the people of Williams Harbour, Labrador, especially Jim and Vera Russell for their help and hospitality. The assistance of the many people in the Ocean Sciences Centre's "lipid lab," especially Jeanette Wells, was invaluable.

In: Resetting the Kitchen Table
Editors: C. C. Parrish et al., pp. 51-74

ISBN 1-60021-236-0
© 2008 Nova Science Publishers, Inc.

Chapter 4

LIFE IN THE FAST FOOD CHAIN: OU SONT LES POISSONS D'ANTAN?

Nigel Haggan[1], Cameron Ainsworth[1], Tony Pitcher[1], U. Rashid Sumaila[2] and Johanna Heymans[1]

[1]Fisheries Centre, The University of British Columbia, Canada
[2]Fisheries Economics Unit, Fisheries Centre,
The University of British Columbia, Canada

ABSTRACT

This chapter reviews the decline in high trophic level or 'table' fish on Canada's Pacific coast in the context of global depletion, the potential extinction of marine species and the economic and social drivers of overfishing. Impacts on Aboriginal and other coastal communities are identified. Whole ecosystem (*Ecosim*) models of northern British Columbia as it was in the 1750s and present-day are used to determine the sustainable food production potential of both past and present systems under different exploitation scenarios. The food production potential of the 2000 and 1750s ecosystems was assessed under four twenty-five-year simulation scenarios. Results indicate that the 1750s system could have sustainably generated over twenty times the current annual food production in northern British Columbia's capture fisheries, while meeting UN food security criteria including cultural appropriateness. The present-day system could sustainably produce six times the fisheries yields extracted today if the fleet was adjusted to meet UN criteria for responsible fishing, and fisheries were conducted optimally to make best use of the existing resources.

One hundred-year simulations of the present day (2000) system incorporating natural climate and ocean regime variability indicate that the existing BC fishing fleet poses a significantly higher risk of seriously depleting and extirpating many species than an alternate fleet structured largely upon UN criteria for responsible fishing.

The chapter concludes with a review of the strengths and weaknesses of Marine Protected Areas and other conservation measures in achieving ecosystem restoration. We draft candidate fishing strategies for northern British Columbia that would increase food production; first, in terms of total available protein, and second, in terms of high-quality tablefish production.

INTRODUCTION

The Whale that wanders round the Pole
Is not a table fish.
You cannot bake or boil him whole
Nor serve him in a dish;

But you may cut his blubber up
And melt it down for oil.
And so replace the colza bean
(A product of the soil).

These facts should all be noted down,
And ruminated on,
By every boy in Oxford town
Who wants to be a Don (Belloc 1896).

What is a "table fish"? Why is a whale not one of them? Which "facts" do the poet exhort every aspiring "Don," or high trophic level scholar, to "ruminate on" or consider?

A "table fish" is, as the name implies, something that one would be proud to serve at a dining table, Atlantic salmon (*Salmo salar*), Pacific salmon (*Oncorhynchus* spp.), halibut (*Hippoglossus stenolepis*), lingcod (*Ophiodon elongatus*), red snapper (*Sebastes ruberrimus*), Atlantic cod (*Gadus morhua*) and so forth. The sense of pride and cultural identity wrapped up in the ability to catch and serve fish and a whole range of seafood is well conveyed by Tirone *et al.* (Chapter 10, this volume). This pride is often expressed by cooking and serving large fish whole, hence Belloc's exclusion of the whale from the canon of table fish.

The Haida people of the Pacific Northwest consider Chinook salmon (*Oncorhynchus tshawytscha*), as "rich food" which was "essential for maintaining the dignity of the family by possession and distribution at potlatches" (Boas 1916, in Jones 1999). Smoked halibut cheeks, called *Xang* in Haida were said to be a special food of chiefs (Jones 1999). The "facts" to be considered are that whalers traveled the globe, at great risk and hardship, to hunt something that was not used for food, at least not by English and American[1] whalers, and could well be replaced by an agricultural product. Why so? It was highly profitable. Lives were lost, but fortunes were made. Whales everywhere were driven to the brink of extinction. Whaling is the classic example of Colin Clark's dispiriting insight that dollars grow faster than long-lived marine species, so it makes economic sense to catch them all and put the money into something that gives a higher rate of return (Clark 1973).

> **Whales and coastal food security**
>
> Whales and other marine mammals are vitally important as food for Inuit and other circumpolar communities and were traditionally hunted by coastal tribes in the Pacific Northwest (Monks 2001 and references therein; Brody 1994). This hunt has been revived in recent years by the Makah people in Washington, although not without controversy. The Haida, Nuu-chah-nulth and possibly other First Nations in BC assert, but have not yet exercised, an Aboriginal right to hunt whales. See Reeves (2002) for a global review of Aboriginal whaling

---

[1] But see 'Stubbs Supper' - Chapter 64 of Moby Dick (Melville 1851).

Mining The Provident Sea[2]

> . . . the cod fishery, the herring fishery, the pilchard fishery, and probably all the great sea fisheries are inexhaustible; that is to say that nothing we do seriously affects the numbers of fish. And any attempt to regulate these fisheries seems consequently, from the nature of the case, to be useless (Huxley 1883).

Just eighty-five years later, Jack Davis, Canada's then Minister of Fisheries, compared the fishery to a copper mine, in which the best ore is taken first before the miner turns to progressively larger volumes of lower grade ore until the mine is exhausted: "Mr. Davis then turned to the sea and explained that life there is built on the same type of pyramid, at the top is the whale and below it such species as the salmon and the tuna. As the base broadens out it contains fish successively smaller but in greater number until, at the bottom, is the limitless mass of plankton which supports the whole pyramid." The Whale, the Minister said, has been virtually wiped out and the tuna and the salmon will be the next to go as man works his way down the pyramid to the plankton (*North Island Gazette* 1968, in Meggs 1991). This process, now known as "fishing down the food web" has been shown to be taking place in Canada and globally (Pauly et al. 1998, 2001).

> **What problem?**
>
> Apart from a handful of fisheries and social scientists and NGOs, the public is blissfully unaware of the extent of depletion. A 2001 survey found that British Columbians believe 16% of their waters to be 'no-take' marine protected areas (Strategic Communications Inc. 2001). Atlantic Canadians believe that 20% of their waters are fully protected (Edge Research 2002). In fact, less than 0.1% is fully protected. Canada's markets are full of fish, increasingly of distant origin, but this is not widely known.
>
> The live fish markets of Hong Kong are bursting with a multitude of species, flown in from increasingly depleted coral reefs in Indonesia and New Guinea, but the only fisheries supported by Hong Kong waters are for prawns, and small, high-turnover pelagic species harvested for agriculture and aquaculture feed. Problems in the westcoast fishery are cast in the language of allocation, not conservation.

Overfishing is now accepted as the major cause of the depletion and changes to marine ecosystem structure (Hilborn et al. 2003; Hall 1999; Christensen et al. 2003). The extent of recent fishery depletions and collapses is even more serious than many had thought (e.g., large fish, Myers and Worm 2003; table fish biomass, Christensen et al. 2003; whales, Roman and Palumbi 2003; sharks, Baum et al. 2002, Schindler et al. 2002; turtles, Hays et al. 2003).

Even exceedingly productive species such as hake (*Merluccius* spp.), deemed to be capable of sustaining intensive and prolonged fisheries (Pitcher and Alheit 1995), have been depleted to the point of closure, provoking a 1999 protest by industrial fishers in Chile that included a port blockade by one hundred fishing vessels and a march on the Chilean Congress (*Reuters* 1999). In a June 2005 interview, Cosme Caracciolo, President of the association of Chilean artisanal fishermen, stated that artisanal fishers have been virtually excluded from

---

[2] See Cushing (1987).

fisheries for jack mackerel (*Trachurus murphyi*) and two species of hake (Soto 2004), both of which go to fishmeal to supply explosive growth in Chilean salmon aquaculture.

Small, highly-productive species, collectively known as 'forage fish' are pivotal in the marine food web as they transport energy from plankton to higher trophic levels. Figure 4.1 shows some of the ecosystem connections to capelin (*Mallotus villosus*) in the Newfoundland and Labrador ecosystem as modelled by Bundy *et al.* (2000).

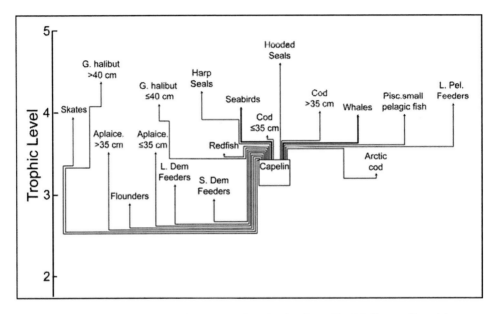

Figure 4.1. Simplified ecosystem model showing pivotal role of capelin (*Mallotus villosus*) in Newfoundland and Labrador (areas 2J3KL). Groups not connected to capelin removed from original figure. Alida Bundy, Fisheries and Oceans, Canada, unpublished data. See also Bundy *et al.* (2000)

Forage fish have long been converted to fishmeal used in pig and chicken feed. Salmon farming now accounts for some 50% of forage fish use in Europe[3]. Fishmeal is used extensively for salmon aquaculture in North America and Chile and for shrimp feed in the Far East (IFFO n.d.; Miller 2003). While some small pelagic species such as horse mackerel and sandlance are unpalatable, others such as herring, capelin, sardines, and anchovies, are eminently suitable for human consumption. Eulachons (*Thaleichthys pacificus*) are of extremely high cultural and economic importance to Aboriginal people in the Pacific Northwest (Drake and Wilson 1991), and are particularly high in lipids which contain healthful omega-3 fatty acids (Turner et al. Chapter 1, this volume).

Role of Sport Fisheries

Sport fisheries are sometimes perceived, and inevitably portrayed, as a minor part of catch, but a major contributor to the economy. Unrecorded and unregulated sport fish catch is a matter of serious, ongoing concern among Aboriginal and commercial fishers in the Pacific

---

[3] Now, after collapse of North Sea herring (ICES 1995) mainly sandlance (*Ammodytes* spp.), as yet unexploited in BC.

Northwest (Pitcher et al. 2002a and references therein). In fact, sport fishing accounts for 12% of global catch of table fish (Cooke and Cowx 2004) an average of 25% in the US and in excess of commercial catch of several species (Coleman et al. 2004). Other adverse effects of sport fisheries include depleting refugia too small to be targeted by commercial fisheries and so depriving Aboriginal people of a vital 'last resort' for food and cultural purposes on a coast where commercial long-lining has depleted the accessible grounds (Figure 4.2). Aboriginal people often regard sport fishing, particularly 'catch and release fisheries' as 'playing with food' which is unethical (Jones and Williams-Davidson 2000). The search for 'trophy' fish, involving catch and release of smaller specimens is both "disrespectful to the fish" and is a form of perverse genetic selection that reduces average size over time. Sportfishing in marine protected areas is particularly adverse as it defeats the major objective of developing a population of large, old, fecund spawners (Birkeland and Dayton 2005, and references therein).

Figure 4.2. Heiltsuk Nation members Pam and Bessie Brown protesting sportfishing expansion in the central coast. Photo: Janet Shaw

Life in the Fast Food Chain

In the ecological sense, life in the fast food chain relates to the replacement of long-lived high trophic level fish with short-lived species (Pauly et al. 1998). This situation is at its most extreme in places like the South China Sea (Cheung and Pitcher 2004; Cheung and Sadovy

2004), but also happens in Canada (Pauly et al. 2001). Intensive commercial and sport fisheries with unlimited access to the ocean also reduce the average age and size of fish caught over time (Law 2000). This is problematic, as many fish live to a great age. Table 4.1 shows the maximum ages, size and weight for salmon and other important table fish from the west and east coasts of Canada. Fish of this age and size would have been rare at any time, but are virtually unknown today.

Recent research indicates that reproductive success is more closely related to fish age than size (Berkeley et al. 2004), and that the number of *effective* spawners may be orders of magnitude lower than the total number of females in a population (Hauser et al. 2002). This underscores the need for extreme caution when setting catch rates for long-lived species. The failure of single species science and management, and the fleet structures that have evolved in response demands an ecosystem approach to restoration. Setting restoration targets requires an exploration of what the ecosystem is capable of producing on a sustained basis. In the social sense, life in the fast food chain relates to the nutritional content, cultural appropriateness and preferences aspects of the World Summit Food Security definition.

Table 4.1. Maximum known age, weight, and size for some Canadian table fish

Common Name	Scientific name	Years	Kg	Cm
Atlantic cod	*Gadus morhua*	25	96	200
Blackcod	*Anoplopoma fimbria*	94	57	120
Pacific Halibut	*Hippoglossus stenolepis*	55	363	267
Lingcod	*Ophiodon elongatus*	33	59.1	152
Red snapper	*Sebastes ruberrimus*	121	17.8	104
Pacific Herring	*Clupea pallasii*	19	1	46
Rougheye rockfish	*Sebastes aleutianus*	205	0.9	97
Pac. Ocean perch	*Sebastes alutus*	98	1.4	51
Copper rockfish	*Sebastes caurinus*	50	2.74	58
Atlantic Salmon	*Salmo salar*	13	46.8	150
Pacific Salmon	*Oncorhynchus*			
Chinook	*O. tshawytscha*	9	61.4	150
Sockeye	*O. nerka*	7	7.71	84
Coho	*O. kisutch*	5	15.2	108
Chum	*O. keta*	6	15.9	100
Pink	*O. gorbuscha*	3	6.8	76
Steelhead	*O. mykiss*	11	25.4	120

Sources: FishBase, *www.fishbase.org* and Alaska Dept. of Fish and Game, *http://tagotoweb.adfg.state. ak.us/ADU/maxagetable.asp*

Issues include the accelerated growth of farmed species from chickens to salmon (Volpe, Chapter 5, this volume), but likely the most significant factor is the increasing rate at which we live our lives and the attention span of small children and youth exposed to cartoons larded with advertising messages of instant gratification. The 'junk food giants' have an increasing presence in North American schools and universities. In 1995, The University of British Columbia signed a ten-year sole-source contract for US$6.8 million with the Coca Cola Company, becoming the first, but by no means the last Canadian university to do so (Thomas 2005). Parents and educators are beginning to resist the presence of junk food in schools and politicians are starting to heed. The UK announced a ban on junk food in schools from September 2006, in large part due to a TV campaign and petition by celebrity chef Jamie Oliver to involve school children in the design and preparation of healthy meals (Naughton 2005). In Canada, junk food has been banned in Ontario schools (Alphonso 2004). British Columbia Education Minister Shirley Bond plans to eliminate junk food over the next four years (Bond 2005).

The advertising pressures are of particular concern to Aboriginal communities, *viz* the ease of pulling food off the supermarket or convenience store shelf compared to the very limited time young people get to spend with elders and family on the land, and the weeks it takes to catch, dry, and prepare traditional foods such as seaweed and eulachon grease (Turner 2005). The cultural and dietary value of eulachons has been noted before, however many Aboriginal children now prefer tomato ketchup (pers. comm. Faren Brown-Walkus, Heiltsuk Nation, age thirteen). The prevalence of junk food and sugary drinks is a major cause of obesity-related disease in Aboriginal communities and the general population (Parrish et al. Chapter 13, this volume; Wong 2004; Turner and Ommer 2004).

Ecological, Economic and Cognitive Drivers of Overfishing

Pitcher (2001) identified ecological, economic, and cognitive processes, or "ratchets" that drive depletion. The ecological and economic ratchets are inextricably linked. As large fish get scarce, fishers buy bigger, more powerful vessels, fishing gear, and high-tech electronics. The capital tied up in Pacific vessels, licences, and quotas is around US$2 billion (Ecotrust 2004; Nelson 2004). Increasing corporate concentration and the need to service this capital is a major economic driver of overfishing (Clark 1973).

Global subsidies of US$20 billion *per annum* (Milazzo 1998) keep fleets active long after fishing has ceased to be economically viable. Even so-called 'green' subsidies such as vessel buybacks have been shown to be counterproductive (Munro and Sumaila 2002). In Canada, subsidies to fishing have been estimated as at least equal to the catch and employment value (Pitcher et al. 2002b and references therein). Estimated overall figures for government subsidies to fishing in Canada range from a 'zero sum game,' where public money equals the value of the catch (100% subsidy, Dr Mary Gregory, Department of Fisheries and Oceans, Ottawa, pers. comm. 1999) to 150% (Roy 1998) to 170% subsidies in Newfoundland prior to the collapse (Schrank et al. 1987).

The cognitive ratchet, or "shifting baseline syndrome" where successive generations perceive the abundance and relative size of fish that existed in their early days as what there ought to be, is subtle and hard to reverse (Pauly 1995). Throughout human evolution, the ocean has been a metaphor for all that is mysterious, abundant, and inexhaustible (Haggan 2000), so, while it is easy, however sad, to contemplate the extinction of the giant panda, it is difficult to comprehend depletion and extinction risk in the ocean.

What is happening in the ocean parallels the extinction of large land animals as people spread over the face of the earth (Diamond 1997), a fate also suffered by large marine species such as Steller's sea cow (*Hydrodamalis gigas*) (Pitcher 2004c). Reduced access to marine protein also contributes to depletion of smaller land animals through increased pressure on 'bushmeat' (Robinson and Bennett 1999; Brashares et al. 2004). The significant difference is that the first wave of terrestrial extinctions took place over millennia, the second wave over a few hundred years of the age of exploration. It is only in the last one hundred years that we have developed the technology to catch all the fish in the sea, and deployed it so successfully that, depending on species, large fish hover between one tenth and one hundredth of their pre-industrial fishery abundance.

First Nations and Food Security

First Nations in the Pacific Northwest traditionally fished across the food web, taking a complete range of resource from whales (Monks 2001) to salmon, to shellfish, to seaweed (Turner et al. Chapter 1, this volume). The high level of social complexity and cultural richness attained by Pacific Northwest tribes has previously been attributed to the variety and year round availability of abundant resources. Recent research indicates that First Nations were not mere passive beneficiaries of this abundance (Anderson 2005; Deur and Turner 2005). Aboriginal people developed the technology to intercept entire salmon runs between 3,000 and 4,000 years ago, yet, according to pre-contact fishing and consumption levels, were able to sustain, enhance, and manage salmon populations to sustain catches equal to or greater than the present commercial fishery (Jones 2002; Haggan et al. 2006), now deemed "unsustainable" (McRae and Pearse 2004).

The sustainable First Nation fisheries would however, have differed significantly from those of the present day in species enhanced and capture technology. Dried salmon was essential for winter food and trade. Chum and Pink salmon keep better than the richer sockeye, Chinook, and coho, so would have been required in large numbers. Most traditional fisheries were 'terminal,' i.e. targeted stocks returning to their stream of origin, as distinct from today's 'interception' fisheries that target migrating stocks in saltwater with the inevitable consequence of overfishing weaker populations (Glavin 1996). The Pacific Northwest may have been a wilderness 4,000 years ago but the 'inexhaustible' resources described by early explorers and settlers were the result of conscious effort. Much can be learned by studying pre-contact management and enhancement methods (Wright 2004; Haggan et al. 2006; Turner 2005). The pre-contact Beothuk and Inuit people of Newfoundland and Labrador also relied heavily on the marine ecosystem (Marshall 1996) and their diet consisted of 65% marine derived food, predominantly seals, salmon, and birds (Heymans 2003). West coast First Nations have been severely impacted through disappearance of food sources readily available for thousands of years (Turner et al., Chapter 1; Ommer et al., Chapter 8, this volume). Some traditional resources can still be accessed, but often with increased effort, cost, and risk to life through, for example, having to go further out to sea in small boats in adverse weather conditions (Richardson and Green 1989).

MODELLING FOOD SECURITY AND EXTINCTION RISK

The Back to the Future Project

The collaborative Back to the Future approach engages natural and social scientists and the maritime community in constructing models of present and past ecosystems (Haggan 2000; Pitcher 1998b, 2000, 2004a, 2005; Ainsworth and Pitcher 2005a). The objective is to build support for whole ecosystem restoration goals that relate to benchmarks of past abundance, diversity, and trophic structure rather than present scarcity. The Back to the Future component of the *Coasts Under Stress* project (Ommer and team, in press) used *Ecopath* (Christensen and Pauly 1992) to model the Newfoundland and southern Labrador[4] ecosystem for the 1450s, 1900s, 1985, and 1995 (Pitcher et al. 2002c; Heymans and Pitcher 2004) and the northern BC ecosystem for the 1750s, 1900s, 1950s, and 2000 (Ainsworth et al. 2002). The northern BC models aggregate ecosystem components into fifty-three 'functional groups' of species having similar diets. *Ecopath* models represent a 'snapshot' of an ecosystem at a particular time. Figure 4.3 represents the proportional decline in biomass from the 1750s to the present day in northern BC from four *Ecopath* models.

Ecosim (Walters et al. 1997) allows us to ask 'what if' questions on the effect of combinations of fishing rates, management, and conservation measures over time periods from twenty-five to one hundred years. *Ecosim* was used to compare the food production capacity of the 1750s and 2000 ecosystems as modelled by Ainsworth *et al.* (2002). For all food security comparisons, the 2000 and 1750 models were subjected to twenty-five years of simulated fishing.

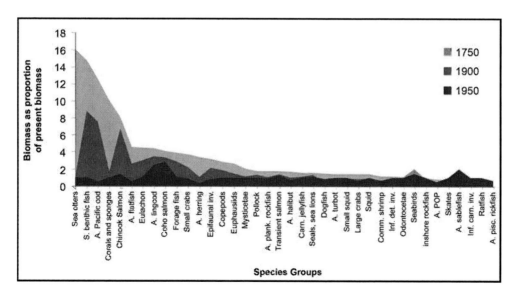

Figure 4.3. 1750s, 1900s, and 1950s northern BC biomass as a proportion of 2000 biomass. As modelled by Ainsworth *et al.* (2002)

---

[4] DFO statistical areas 2J3KLNO.

Optimal fishing mortalities were estimated by *Ecosim* using the policy search routine of Christensen and Walters (2004)[5]. The optimal plans describe the gear and effort configurations that will yield the maximum amount of protein in tonnes per year. The optimization settings do not consider catch composition; i.e., all species groups are assumed to be of equal desirability and/or nutritional value. Later, we discuss the implications of this assumption in terms of UN food security preference and cultural-appropriateness criteria.

Protein yield was assessed using vessels and fishing gear in current use in northern BC and an alternate configuration, or *Lost Valley*[6] scenario, a metaphor for a pristine ecosystem, untouched by human fishing (Pitcher et al. 2002d, Pitcher 2004b). Simulated fisheries are used to determine the productive potential of the pristine system or Lost Valley. Fishing gears used in Lost Valley simulations are derived largely from the UN Code of Conduct for Responsible Fishing (FAO 1995) which Canada played a lead role in developing. The Lost Valley has been used in numerous simulations for northern BC (Pitcher and Ainsworth, in press; Ainsworth and Pitcher, in press; Pitcher et al. 2005; Ainsworth and Pitcher 2005b), and elsewhere (Newfoundland: Pitcher et al. 2002d; Hong Kong: Pitcher et al. 2002e).

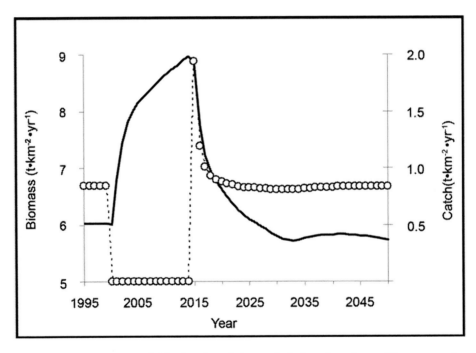

Figure 4.4. Tablefish biomass (solid line) and catch (open circles) projected over fifty years under restoration scenario for 2000 ecosystem. Fishing ceases (2000–14) allowing tablefish biomass to rebuild. Fishing resumes in 2015 using present-day fishing mortalities. The restored biomass is lost in only three to four years

---

[5] As the policy search routine has no explicit option to maximize catch, a simplified economic analysis was used to determine optimal fishing rates for food production where all fishery products have equal value, regardless of species caught or fishing gear used. By default, the economic optimization maximizes net present value, but by setting a discount rate near zero, the policy search routine can be made to maximize equilibrium level harvests.

[6] The reference is to a fictional scenario where explorers find themselves in a 'Lost World' where creatures from time past live untouched by human influence (Doyle 1912).

The Lost Valley scenario would be recognizable by today's fishers as it retains many of the fishing gears currently in use. It differs in that it reduces the number of juvenile fish groups caught and reduces bycatch to levels practically achievable through gear modification. It also assumes that bottom trawls for groundfish and shrimp have been modified, within realistic limits, to prevent damage to corals, sponges, and other seabed structure (Ardron 2005). The level of fishing effort is a key consideration in the scenarios, and is critically important for ecosystem restoration. Figure 4.4 shows the results of a simple rebuilding plan for the 2000 ecosystem. The beneficial effect of a fourteen-year program to rebuild the ecosystem by shutting off all fishing would be negated in only three to four years of renewed fishing at current fishing levels. This realization is at the root of our exploration of alternate gear and effort configurations in northern BC.

RESULTS

Current Gear–Protein Maximization Scenario

The *Current Gear–Protein Maximization* scenario uses the seventeen major fishing gears used by the northern BC fishing fleet to exploit twenty-five of the fifty-three functional groups in the year 2000 model. This scenario predicts that after twenty-five years of simulated fishing, the 2000 ecosystem could sustain more than four times the current landings. In this case, the ecosystem will have been restructured through selective manipulation and cultivation of species to provide increased levels of catch, even though for most fisheries the optimal fishing effort will be lower than today's levels.

Applied to the 1750s model, the Current Gear–Protein Maximization scenario was able to sustain catches greater than nine times the current level, although fishing effort per gear was adjusted optimally. Fleet configuration, catch composition, bycatch, and discard rates were entered in the same proportion as in today's fisheries[7]. This is for illustrative purposes only, as Figure 4.4 indicates today's fishing effort level would be inappropriate, even in the much more abundant ecosystem of the 1750s.

Current Gear–Economic Maximization Scenario

The *Economic Maximization* scenario considers prices for species, the cost of fishing and discount rate. The Economic max scenario increases landings to 3.7 times the current production rate in the 2000 model and 7.5 times for the 1750s. While this is less than the protein maximization scenario, it improves the quality of catch and so reflects consumer

---

[7] Some functional groups have increased in productivity since pre-contact as a consequence of exploitation, since older, less productive individuals are removed from the population. This and other factors, caused current catches, when transposed directly, to throw the 1750 model out of balance. Therefore, the absolute level of fishing effort was reduced by an equal proportion across all functional groups, to about 10% of its overall original value as seen in the 2000 model. This adjustment will not affect the results of the policy optimization; it is only a modelling convenience. Since the policy search was free to vary relative fishing mortalities from year zero of the simulation, the initial (reduced) catch was immediately discarded in favour of an optimal gear deployment solution.

preference (e.g. more table fish, less small fish or invertebrates); in terms of total protein production, the economic maximum still provides a great improvement over current BC fisheries.

Ecological Limit Scenario

In the real world, few fishing gears are entirely selective in the species they catch. Total fishing mortality comprises: catch of target species; 'bycatch' of other commercial species for which the fishers are not licensed; 'discards' of small, immature, and non-commercial species; and species that may be damaged by gear, or have their behaviour altered in a way that makes them more accessible to predators. Atlantic cod have been subject to a moratorium since 1992, but bycatch levels of cod in other fisheries are such that they are unlikely to recover (Rosenberg et al. 2005).

Mortality of non-target species further reduces the productive potential of the ecosystem, for example, when structure-building organisms such as corals and sponges are removed (Ardron 2005). In an ecosystem context, simultaneous capture of multiple species has other, often-overlooked effects on the productive capacity of the ecosystem through complex foodweb interactions (see also Figure 4.1).

Gear that catches multiple species also limits our ability to manipulate the ecosystem. We cannot easily choose to rebuild one weak species, while increasing catch on a sympatric species using unselective gear. If we could remove the distortions imposed by licensing and regulatory systems and the limitation imposed by the technology of fishing, we would uncover the true ecological limit of production. This limit could be used as a benchmark to evaluate the effectiveness and compatibility of a given suite of fishing gears.

> **Ecological or Social Risk?**
>
> Restricting catch to protect weaker stocks and 'non-commercial' species is always contentious. In 2004, the Committee on the Status of Endangered Wildlife in Canada (COSEWIC) recommended listing the Cultus and Sakinaw sockeye salmon under Canada's Species at Risk Act. The Environment Minister overruled the recommendation on the basis that these populations were 'a fraction of 1%' of total BC sockeye and that their protection would incur 'unacceptably high social and economic costs' (Canada 2004).

The *Ecological Limit* scenario establishes a theoretical ecological limit on food production for both the 2000 and 1750 ecosystems in the absence of technological or regulatory constraints. Each species is assigned to a dedicated anonymous fishing gear to determine a hypothetical "ecologically sustainable yield" obtainable if fisheries were 100% 'clean,' i.e. had no bycatch or discards. Hence, the policy search routine is able to independently vary fishing mortality among species, and sculpt the ecosystem into a configuration that satisfies our objective (Ainsworth et al. 2004). The Ecological Limit scenario is constrained in that it has access only to the functional groups that are exploited by present-day BC fisheries.

The Ecological Limit scenario increases landings to six times the current production rate for the 2000 model and eleven times for 1750. This scenario allows us to set an upper limit to productivity as determined by the ecology of the system.

The two following scenarios investigate the hypothesis that increasing the variety of species landed will augment total food production from the ecosystem. We expect this to be achieved by two means. First and simply, there is some potential to increase fisheries landings—particularly on invertebrates and deep-water forage fish (which are made available by contemporary fishing technology) and seals, which were once of considerable importance in the diet of First Nations in northern British Columbia (Monks 2001, and references therein). These populations could sustain new fisheries and contribute to overall food production. Secondly, as more ecosystems become subject to direct population control by fishing, it becomes easier to manipulate the ecosystem into a desired configuration, in this case maximum protein production.

Lost Valley Scenario

The Lost Valley differs from Current Gear scenarios in that it revives efficient and selective Aboriginal fishing technologies such as traps, weirs, and fishwheels. Another significant difference is that it expands the number of functional groups exploited from twenty-five to thirty. The 'new' groups include those that are of traditional importance to First Nations, such as marine mammals and invertebrates. The Lost Valley scenario also includes some emerging and potential fisheries such as live rockfish (*Sebastes* spp.), squid, and jellyfish. This configuration is somewhat idealized, containing only minimal bycatch and discards, but species catch compositions are realistic for each fishing gear modeled. It therefore represents the level of food production that may be realistically achievable through regulation and gear modification.

Increasing the number of functional groups fished to thirty allows the Lost Valley scenario to augment current food production by eight times in the 2000 ecosystem, and twenty-six times in the pre-contact ecosystem. This is achieved by the sustainable exploitation of large invertebrate populations (primarily bivalves, sea urchins, and sea cucumbers) whose cumulative biomasses in the 1750s were estimated to be as much as four times the present-day levels (Ainsworth et al. 2002).

Lost Valley Broad Scenario

To understand the relationship between available food production and the range of target species, we have expanded the Lost Valley scenario to include some additional target species. The *Lost Valley Broad* scenario includes fisheries for baleen whales (which is of consequence only in the 1750 comparisons), additional species of forage fish and toothed whales[8]. This scenario exploits thirty-four functional groups.

With more functional groups subject to exploitation, fewer ecosystem components are beyond the direct control of the policy search. The Lost Valley Broad scenario improves food production by approximately eight times current levels in the 2000 system and twenty-eight

---

[8] Orca, dolphins and porpoises. Note that the Lost Valley Broad scenario is designed to explore how much protein the ecosystem could produce, not to recommend fisheries on these or other species. Orcas are of very high cultural significance for BC Aboriginal people, so are unlikely to have been hunted.

times in the 1750 system. However, this is only a marginal improvement in food production over the previous Lost Valley scenario which already exploits a large proportion of ecosystem components.

IMPLICATIONS FOR TABLE FISH

Up to a certain point, increasing the number of species approaches fishing *across* the food web by maintaining trophic relationships in a proportional sense, as opposed to fishing *down* (Pauly et al. 2001, 1998). Past that point, extreme protein production scenarios, such as the Lost Valley and Lost Valley Broad, increase the production of low trophic level species like invertebrates at the expense of table fish. Figure 4.5 shows the results of optimizing for protein production. After twenty five years of manipulation, the 2000 ecosystem could sustain much greater catch rates than are currently realized by BC fisheries. The 1750 ecosystem would have been able to sustain greater catch rates still than the 2000 ecosystem. Fisheries for table fish continue to some extent in these extreme food production scenarios, but cannot hope to match the quantity of protein from invertebrates due to limits imposed by the carrying capacity of the ecosystem.

Sustainable Catches after Twenty-Five Years

Table 4.2 presents sustainable catch rates after twenty-five years of simulated fishing for the five scenarios as multiples of current annual catch. With reduced effort, the current BC fleet structure is able to manipulate the ecosystem to support much greater catch rates than are currently enjoyed once the priority for protein production is factored into the long-term policy agenda. The ecological limit on protein production from currently-targeted species is revealed by the Ecological Limit scenario. Increasing the number of species, as in the Lost Valley and Lost Valley Broad scenarios, generates a marked increase in food production over current levels. Regardless of scenario, the pre-European contact ecosystem vastly outperforms the present-day ecosystem, indicating that sustainable production potential has been lost.

Table 4.2. Potential protein production rates for five fishing scenarios in northern BC

Scenario	# of groups fished	Protein production rate after 25 years relative to current BC fisheries	
		1750 model	2000 model
Current Gear (protein max)	25	9.1	4.1
Current Gear (economic max)	25	7.5	3.7
Ecological Limit	25	11	5.7
Lost Valley	30	26	8.0
Lost Valley Broad	34	28	8.4

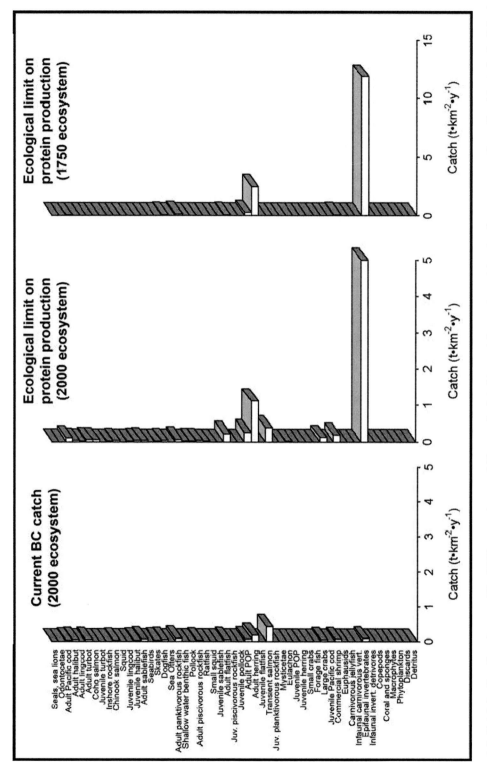

Figure 4.5. Catch by functional group. The 2000 ecosystem could be manipulated to sustain greater catch rates than are currently realized by today's fisheries. The 1750 ecosystem was capable of producing much more. Protein production is maximized using large invertebrate fisheries

Figure 4.6 summarizes the sustainable fishing capacity of the 2000 and 1750s ecosystems under different extraction scenarios. Bar graphs show the equilibrium (end-state) rate of production in terms of annual catch for pelagic or surface-living fish, demersal or bottom-dwelling fish and invertebrate functional groups. As a baseline for comparison with the theoretical optima, the year 2000 food production in northern BC shown in the left-most bar in Figure 4.6a is derived directly from catch statistics (see Ainsworth et al. 2002 for data sources).

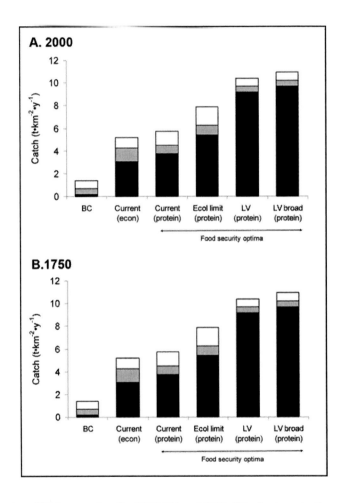

Figure 4.6. Optimal equilibrium catches for 2000 (A) and 1750 (B) after twenty-five years. Black bars show invertebrates; grey bars show demersal fish; white bars show pelagic fish. Scenarios left to right: 'BC' shows real-world catch in northern British Columbia in 2000; current gear types optimized for economic performance; current gear types optimized for protein production; Ecological limit on protein production from currently targeted species; LV (Lost valley) pursues additional fish species; LV broad pursues additional fish and invertebrate fisheries

The economic objective (Econ) shows the equilibrium catch level when fishing is optimized for maximum economic return over twenty-five years, as opposed to catch with the food security optima, also optimized over twenty-five years. Value is a fair representation of *"preference,"* in the UN food security definition to the extent that it reflects what customers are prepared to offer.

It does not, however, reflect all species used by First Nations or all that could be exploited. Although the point of the economic optimization is not strictly to increase catch, there is still a great improvement over current landings since economic return is linked to catch volume. The four right-most bars in Figure 4.6a and b show food production under the different modelling scenarios.

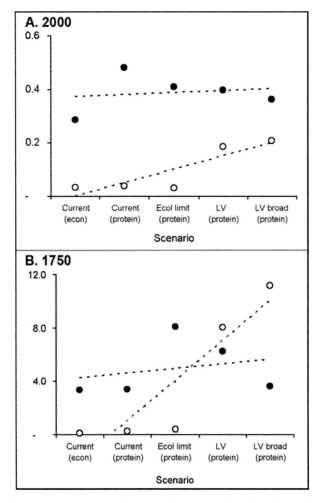

Figure 4.7. Changes in ecosystem structure (average change per functional group) after twenty-five years for 2000 (A) and 1750 (B) ecosystems. Closed circles show changes in exploited functional groups; open circles show changes in unexploited groups. From left to right, increased catch influences the ecosystem more heavily. As ecosystem manipulation increases, unexploited groups are altered from their original condition by more than exploited groups, suggesting that they are manipulated to support higher catch rates. The 1750 ecosystem has a greater potential for manipulation than 2000, both in terms of an absolute change from the initial condition, and in the relative contribution of non-exploited groups towards supporting greater catches. Scenarios left to right: Current gear types optimized for economic performance; current gear types optimized for protein production; Ecological limit on protein production from currently targeted species; LV (Lost Valley) pursues additional fish species; LV broad pursues additional fish and invertebrate fisheries

Whole-Ecosystem Manipulation

Figure 4.7 shows evidence of whole ecosystem manipulations by the optimal fishing policies in order to support a larger sustainable catch.

Relative biomass changes from the initial ecosystem condition (as a result of optimal fishing scenarios) are compared between functional groups which are subject to fishing, and those which are not. After twenty five years, non-exploited groups have changed from their original condition by more than those subject to fishing, indicating that the non-exploited groups have been maneuvered to support the additional food production through trophic interactions. This may be viewed as a type of 'ecosystem service.'

From left to right in Figure 4.7, as the fishing strategy becomes more comprehensive, as in the Ecological Limit scenario, or with a greater breadth of target species as with the Lost Valley and Lost Valley Broad scenarios), two things occur. The sustainable catch increases (see Figure 4.5), and the ecosystem biomass equilibrium are manipulated further from its original condition. Results suggest that the 1750 ecosystem has a greater potential for manipulation than 2000, both in terms of an absolute change from the initial condition and in the relative contribution of non-exploited groups towards supporting the fished groups.

EXTINCTION RISK POSED BY CURRENT FLEET *VS* ALTERNATE STRATEGIES

One man stood before the microphone, his face grey with fatigue and anxiety, and said in a breaking voice: "Let's face it: we've caught them all" (Storey 1993, in Ommer 1994).

The words of a fisher at a post-mortem on the collapse of the Atlantic cod, one of the great fish stocks once deemed inexhaustible (Byron 1812–18; Huxley 1883) signal the dawn of awareness that we can drive fish populations to the verge of biological extinction. This realization is also recent in the scientific community (Pitcher 1998a, 2004c; Carlton et al. 1999; Cheung and Pitcher 2004).

The depletion/extirpation risk posed by the Current Gear *vs* the Lost Valley scenarios was determined by driving the 2000 model for one hundred years with a random selection of primary production multipliers derived from an annual climate reconstruction series from 1638–1988 (Gedalof and Smith 2001). Figure 4.8 thus simulates the combined effect of fishing and *natural regime variability* over 350 years ending 1988.

Figure 4.8a shows that euphausiids, copepods, and squid are at significant risk of 80% depletion under either configuration. Lingcod are also at high risk, but substantially less in the Lost Valley scenario. The risk of extinction shown in Figure 4.8b is negligible or substantially less for the Lost Valley configuration than for the Current Gear scenario. The results indicate that fishing the present day ecosystem with fully sustainable and responsible fisheries considerably reduces the risk of extinction and extirpation in the face of past natural climate variability. Climate change may increase natural variability in the future, but has already been shown to have the effect of moving the spawning area of fish northwards, as for small demersal fish in the North Sea (Perry et al. 2005) and herring in the Strait of Georgia (Dr Tom Therriault, Fisheries and Oceans Canada, pers. comm.). A relatively small rise in ocean temperature could make BC waters unsuitable for sockeye salmon (Welch et al. 1998) with devastating effect on Aboriginal people and the commercial fishery.

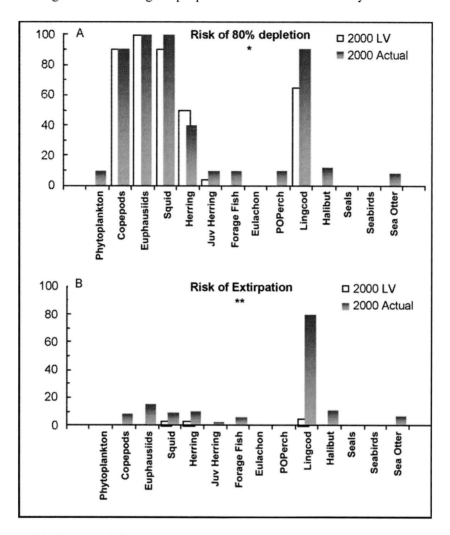

Figure 4.8. Risk of 80% depletion (A) and extirpation (B) in the year 2000 northern BC ecosystem with climate variation and model parameter uncertainty after one hundred years of fishing with the 'Actual' (present-day) and Lost Valley (LV) fisheries. Risks are estimated from one hundred Monte Carlo simulations with all principal model parameters allowed to vary within a 40% range. Risks are compared using the Wilcoxon paired signed ranks test: * = significant at the 5% level; ** = significant at the 1% level

RESTORATION—IMPEDIMENTS AND INCENTIVES

However discomforting as an audit of our management performance, the past models show significant potential for restoration (Figure 4.3; Table 4.2). The downside is that achieving the benefits would require a sharp reduction in catch, with short- to medium-term costs to industry. Single species assessment and management is problematic because it creates divisions within fisheries science and management that are reflected in the structure of government and industry (Finlayson 1994). Hence, the current BC fleet configuration is a product of 150 years of perverse evolutionary pressures, most notably reduction in number of species *per* fishing licence, leading to a proliferation of vessels, licences, and quota valued, at ~US$2 billion (Ecotrust 2004; Nelson 2004). Escalating costs of fishing and buyback or 'fleet rationalization' plans entrain loss of licences by small vessels with devastating effects on coastal communities and small-scale fishers (Gislason et al. 1996).

Individual fishers and their families have heavy, often multi-generational, investments in knowledge and skills, vessels, fishing gear, and social relationships ranging from collegial industry associations to struggles with management (Newell and Ommer 1999, and references therein). This investment relates almost exclusively to the narrow range of species fished. The need to protect these economic, cultural, and social investments escalates as fish stocks are depleted, perpetuating scientific, management, and industry structures that are extremely resistant to change (Finlayson 1994). Increased specialization leads to reduced ecosystem knowledge and higher competition.

On the incentive side, simulated rebuilding to 1950s biomass levels shows that moderate restoration trajectories can give a positive return equal to or greater than bank interest at 5% (Ainsworth and Pitcher, in press.). A new method of quantifying benefits to future generations through intergenerational discounting (Ainsworth and Sumaila 2005; Sumaila 2004; Sumaila and Walters 2005) adds significantly to net present value and strengthens the argument for reinvestment in 'natural capital.'

The first step is to involve the entire maritime community, First Nations, commercial and sport fishers, conservation organizations, managers, and policy-makers in developing management and restoration strategies that increase their collective understanding of the ecosystem. Second, is for government to open the door to radically different approaches that encourage all concerned to harness their creative energy to making it happen.

Can Biodiversity, Table Fish and Fishers Co-Exist?

Tradeoff analysis of restored systems (Ainsworth and Pitcher, in press) shows that maximizing for economic return has a catastrophic effect on abundance, biodiversity, and trophic structure (and indeed seabed structure) in marine ecosystems. This is, in fact, the experiment we have performed over the past one hundred years, driving high trophic level species down to between 10% and 1% of their 1900 biomass globally and on both coasts of Canada. Reduction in fisheries and indeed forestry jobs and revenue is a major driver of the search for 'alternatives' such as farmed salmon and oil and gas. Maximizing for social benefits also has a negative effect on ecological values, while maximizing for biodiversity and ecosystem integrity provides unacceptably low economic and social returns (Ainsworth and Pitcher, in press). Is there any solution that would be a win for biodiversity, table fish, and fishers?

Marine Protected Areas and Fisheries Management

Marine protected areas (MPAs) are increasingly advocated as a way to offset the depletion of marine ecosystems. They are highly effective for the protection of sessile and territorial species, but much less so for moderately migratory species such as cod. Guénette *et al.* (2000) showed that 80% total closure of the east coast fishery would have delayed, but not prevented the cod collapse, while only 20% closure, plus fishing restrictions on migration corridors, would have been effective in averting the collapse. Taken individually, the benefit of MPAs is limited and offset by the contention caused, but a network of MPAs providing protection for inshore areas of high "conservation utility" (Ardron 2002), key oceanic habitat (Worm et al. 2003), and migration corridors (Guénette et al. 2000), would go a long way to meeting biodiversity criteria, preventing depletion and extinction and providing spillover benefits to fishers and ecotourism operators and many other ecosystem values (Sumaila et al., in press). Involvement of the maritime community in the design of such a network is essential to agreement on utility, location, and compliance.

The Role of Quotas

Individual transferable quotas (ITQs) are often advocated as another cure-all for whatever ails fisheries and marine systems. McRae and Pearse (2004) recommend that BC salmon licences be converted to quotas that can be traded on the market. Transferable quotas have two major downsides. Firstly, they lend themselves to concentration in the hands of wealthy individuals and corporations, effectively alienating access to fisheries from Aboriginal and coastal communities, with serious social and cultural consequences including the loss of traditional and local ecological knowledge vital to our understanding of ecosystem function (Coward et al. 2000). Secondly, ITQ holders are only interested in the species they catch, i.e., have no incentive to protect aspects of the system that provide broader cultural, social, ecological, and ecosystem service benefits (Sumaila and Bawumia 2000).

Community quotas, defined as the permanent vesting of access rights in coastal communities, do have the potential to re-link human communities to the ecosystems that called them into being and sustained them for thousands of years in the case of First Nations and hundreds of years for east coast fishing communities (Haggan and Brown 2003).

Policy Implications

The policy implications of ecosystem restoration and greater access for Aboriginal and other communities whose long-term survival depends on ecosystem health include:

- Potential to satisfy Canadian legislative requirements to manage for the benefit of future as well as present generations (Fisheries Act, Oceans Act, Marine Conservation Areas Act) and live up to Canada's obligations as a signatory to the Convention on Biological Diversity;
- Applying intergenerational discounting (Sumaila and Walters 2005) would go a long way towards meeting the First Nation's ethic of seventh-generational thinking;
- Improved conservation and increased food security on both coasts;
- More fish for the settlement of treaties with First Nations and more viable coastal economies with reduced government transfer payments (Haggan and Brown 2003);

- Retention and growth in practice and intergenerational transfer of traditional and local knowledge (Berkes 1999; Berkes and Turner 2006; Garibaldi and Turner 2004; Turner 2003; Turner and Berkes 2006; Turner et al. 2000) with significant benefit to resource management;
- Increased availability of culturally-appropriate and nutritionally-superior foods from local sources, coupled with increased awareness of the health hazards of junk food, as is happening in UK and Canadian schools, would materially contribute to health in coastal communities (Wong 2004); and,
- Strengthen cultural and social activities based on the marine ecosystem and environment.

CONCLUSION

Humans have a surprising ability to change marine ecosystem structure. Evidence suggests that the west coast ecosystem had been manipulated to produce very large surpluses of salmon, invertebrate and terrestrial species long before European contact (Haggan et al. 2006; Turner 2005). Benchmarks of abundance, biodiversity, and trophic structure established by collaborative modelling of past ecosystems indicate substantial potential for restoration (Figure 4.3).

The implications for food security are intriguing. Scenarios that maximize the amount of edible protein divert primary production from table fish to species that do not meet the cultural preferences of Aboriginal and other coastal communities or satisfy the desire of most people to be able to serve and consume large fish such as salmon and cod. These scenarios do not satisfy UN food security stipulations, that food be, *personally acceptable and culturally appropriate . . . produced in ways that are environmentally sound and socially just.*" The criteria of "ecosystem justice" (Brunk and Dunham 2000) are not satisfied, nor are the 'existence' and other values of long-lived, high trophic level species maintained (Sumaila et al., in press).

Somewhat surprisingly, the 'Current Gear–Economic Maximization' scenario that adjusts fishing effort to maximize economic return; came closest to matching UN food security criteria. This is because market price reflects preference for table fish, though we note that not all species traditionally-harvested or valued by Aboriginal people were included in the model.

The Ecological Limit scenario sets a theoretical upper limit on the amount of protein that could be produced by 100% clean fisheries, i.e. with no bycatch, discards or habitat damage. There is significant potential for fishers and other maritime community interests to use this type of scenario to explore alternate ways to catch or otherwise use ecosystem goods and services sustainably. Creating the enabling conditions requires a commitment by government to 'open up' the entire structure of fisheries science and management and to deal effectively and fairly with the ~US$2 billion in vessels, gear, and quota (Ecotrust 2004, Nelson 2004).

In conclusion, ecosystem restoration requires a systemic approach that engages all concerned in the design of MPA networks that protect critical habitat *and* maximize fishery benefits, regulations to protect migratory species and some form of area licensing that provides guaranteed access to communities with a long term interest in ecosystem health and diversity as the key to their own survival. This requires nothing less than a complete overhaul of fisheries science and management matched by government willingness to actively encourage fishers in ways to use their skill to exploit a range of species. Experience with the Back to the Future project indicates that the fishers are willing.

In: Resetting the Kitchen Table
Editors: C. C. Parrish et al., pp. 75-86

ISBN 1-60021-236-0
© 2008 Nova Science Publishers, Inc.

Chapter 5

"SALMON SOVEREIGNTY" AND THE DILEMMA OF INTENSIVE ATLANTIC SALMON AQUACULTURE DEVELOPMENT IN BRITISH COLUMBIA

John P. Volpe

Environmental Studies, University of Victoria, Canada

He who has bread has many problems. He who has no bread has only one problem.
—Byzantine proverb

ABSTRACT

Agricultural industrialization and globalization are often identified as key processes in the erosion of food security and sovereignty. Global seafood production is dominated by modern day hunter-gatherers and so food sovereignty debates until recently have been confined to terrestrial systems. The recent introduction and subsequent spectacular proliferation of industrial marine aquaculture has brought the food sovereignty debate to the sea. The current hub of North American marine-based industrial aquaculture is British Columbia (BC) where salmon farming has grown to become a major factor in shaping coastal communities. Since its introduction into BC in the mid-1980s, the global industrial salmon farming industry has changed dramatically. The meteoric rise of farm salmon production exported from Norway, Chile, and Scotland has caused global salmon prices to plummet, forcing intense industry-wide rationalization. This is typically characterized by large multinationals cannibalizing smaller producers in order to establish economies of scale to counterbalance market price erosion of their product. As a result, global market forces have all but eliminated the sovereignty of both BC salmon farmers and salmon fishermen. Salmon farmers and fishermen—and their communities—must adopt ever-greater economies of scale to remain competitive with growing/consolidating foreign farm producers. All ecological issues associated with farms (organic wastes, parasites, toxins use, escapes etc.) are density dependant and therefore escalation of production results in incrementally greater ecological impacts. Further, the (d-)evolution of salmon from a culturally iconic seasonal delicacy and cultural keystone species of indigenous communities to a low-value global commodity has manifested a broad array of impacts highlighting the codependence of ecology, culture, and economics in coastal communities. Finally, increasing production increases supply, which in turn lowers price and alters consumer expectations of price/value relationships. This not only forces another round of production increases/consolidation on the farm, but also drives out commercial fishermen, particularly smaller owner/operators, who have no capacity to adopt economies of scale in the face of falling salmon prices. The ensuing rationalization process results in corporate fleets, the only model capable of adapting to the economic realities of the commoditization of salmon. Such loss of sovereignty effectively liquidates social capital of coastal communities which is instead invested into corporate machinery paying a lower rate of return far removed from the communities who are now left to pay off the interest as a legacy to the venture.

INTRODUCTION

The recent surge of salmon farming along the coast of British Columbia represents an enormous restructuring of economy and environment that has had major implications for coastal communities. By way of setting context, we begin at *The World Forum on Food Sovereignty* held in Rome in June 2002 in the shadow of the much larger, production oriented, governments-only *World Food Summit*. *The Forum* provided a venue for citizens to extend discussions from simple production strategies, to encompass the rights of producers. Thus *The Forum* was a venue for the voice of the producer and put the concept of "food sovereignty" on the world stage; in retrospect it is now seen as a watershed event in how we address global food politics.

Why this event is viewed with such significance is revealed in its name which invokes "food sovereignty" instead of the more familiar "food security." More than 200 definitions of "food security" could be found in the literature just over a decade ago (Maxwell and Smith 1992). In all cases the right to access and consume food is emphasized. Food sovereignty balances the rights of people to consume healthful food with protecting the autonomy of [often oppressed or marginalized] food producing groups and their rights to decide how to produce food and how to sustain their environments and livelihoods. Food producers are in fact the majority of the world's population and, ironically, the overwhelming majority of the world's hungry (Roberts 2003). Numerous examples exist where the rights of producers and consumers have become unbalanced, particularly where states are obliged to feed all citizens but lack the means to develop and protect the capacity to grow food. For instance, during the Irish potato famine of 1845–46, the amount of food Ireland exported to Britain would have been sufficient to feed those who starved. The 1974 Bangladesh famine resulted from people displaced from employment by floods, leading to widespread economic breakdown—even though more food was produced in the country that year than those years immediately before or after (Sen 1981). These cases illustrate a more general phenomenon; hunger is often a result of inadequate access to, not lack of food.

This is not to suggest the supply of food has always exceeded demand, nor that it always will. In the middle of the last century many populations around the world were certainly food deficient. However this had little apparent effect in stemming population increase. The situation became acute when food demands rose sharply in the baby boom years following World War II when most arable land was already under culture. An array of technical advances (artificial fertilizers, pest-/herbicides, irrigation, artificial selection) combined to produce high yield crop strains genetically capable of responding to the new production model of intense agrichemical and irrigation input. These, combined with adoption of increased mechanization and energy use on the farm, resulted in unprecedented increases of production per acre. When viewed together these technologies summed to enable the large-scale conversion of oil and water into grain. This so called *Green Revolution* ushered in the age of industrial agriculture which succeeded in meeting the demands of the day and has supported the resultant growth of the human population to the present even though the number of farmers working the land in the developed world continues to rapidly decline.

Production capacity of these now conventional technologies, however, is being exhausted and history is poised to repeat as the global human population finds itself on the brink of food deficiency once again. Production per acre has plateaued and in many cases declined, reflecting degradation of the land's production capacity by Green Revolution industrialization (Tilman et al. 2002). The agribusiness funded scientific community has predictably responded by suggesting a new generation of advances—e.g. GMOs (genetically modified organisms)—must be implemented if widespread famine is to be avoided (U.S. National Academy of Sciences 2000). What goes unreported in such techno-centric analyses is that presently global agriculture produces more than adequate supplies to feed every person on Earth a 2,500-calorie daily ration. Despite this, 60 million people die annually as a direct result of insufficient access to food. To put a finer point on the troubling ethical corollary this raises; the same number of people could be adequately fed, thus ending the crisis, by the grain saved if Americans reduced their intake of meat by 10% (Schwartz 2001). Today, as in the past, the critical issue for much of the world's poor remains access to otherwise abundant food, not a shortfall of production. Nonetheless, the complex and nuanced drivers responsible for decreasing global food security and sovereignty are often ignored, resulting in simplistic trickle down prescriptions: hungry people will be fed if we can just produce more food *ipso facto* GMOs are the silver bullet.

Along with assorted techno-fixes such as GMOs, aquaculture—the farming of marine and freshwater species—is now being promoted as a potential solution to meet the perceived shortfall in demand for protein. Aquaculture has an ancient history beginning in China during the Zhou Dynasty, approximately 2,300 years ago (Li 1994). Within 200 years (Han Dynasty) the practice of polyculture (simultaneous culture of multiple, synergistic species) had become widely established and remains a fixture to this day in China and across much of the world. In contrast, modern industrial aquaculture is characterized as being reliant on external inputs of energy towards the production of an export crop with a primary objective of generating profit not food. However, the potential protein production capacity of aquaculture cannot be ignored and many are now calling for a *Blue Revolution*—application of new technologies in order to repeat, in the world's oceans, the "successes" of the Green Revolution on land. Proponents cite three major objectives of the Blue Revolution: to increase food security for those currently or soon to be in deficit; to relieve harvest pressure on wild stocks; and to provide economic opportunity for struggling communities (Skladany and Harris 1995). Deployment of the Blue Revolution in the developed world is based exclusively on intensive production models favouring salmon and shrimp operations over alternative, more sustainable lower trophic level species and more balanced production models such as polyculture (Skladany and Harris 1995).

The Blue Revolution has clearly begun, but it differs from its predecessor, the Green Revolution, in two important ways. First, the Green Revolution's success was based on innovations that enabled increased production without increasing the spatial footprint of operation. While some increases in efficiency in the Blue Revolution are evident, the majority of its increasing production is due to simply increasing the physical area used for operations. Second, the Green Revolution focused on grain (primarily carbohydrate) production, the backbone of the human diet. In contrast, protein is the product of the Blue Revolution. Protein is obviously an important dietary component, however it is very much secondary to carbohydrates in terms of feeding food deficient populations.

The dominant case history that frames the development of the Blue Revolution is salmon farming. Begun in Norway in the 1970s, farmed (Atlantic) salmon has grown to be a dominant global aquaculture product. This process was not without its setbacks. After a decade of meteoric growth, the effects of the industry began to be seen in Norwegian wild salmon populations. The detrimental effects of genetic introgression via farm-wild hybridization have been well documented (Fleming et al. 2000). However the most dramatic effects have been those involving disease and parasite transfers among wild and farmed fish. One of the most alarming events occurred in the early 1980s when farm-borne epidemics of the potentially fatal parasite *Gyrodactylus* devastated wild fish populations. Fuelled by public outrage, the ensuing Norwegian government crackdown on industry resulted in its major reconfiguration. Many operators adapted to the regulations, while others began looking for alternative venues for their operations.

Concurrently, in September 1984 Brian Mulroney was elected Prime Minister of Canada. High on his agenda was the replacement of the Foreign Investment Review Act with the Investment Canada Act. The new legislation essentially removed the necessity that Canadians citizens hold majority ownership of Canadian registered companies.

And so the stage was set for the migration of Norwegian salmon farm companies to Canada. At home in Europe, Norwegian companies were being compelled to conform to strict and in some cases costly new operating procedures, whereas the Canadian federal government threw open the door for them to operate unfettered in Canada. The British Columbia coastline provided exemplary physical and biological habitat required by industry, and regulators were positively predisposed to the arrival of the Norwegians to help solidify the fledgling BC salmon farming industry, at that time based exclusively on farming Pacific (*Oncorhynchus*) salmon species. The influx of Norwegian operators also meant the importation of the Norwegian species of choice—Atlantic salmon (*Salmo salar*).

> We are very strict about the quality and the environmental questions [in Norway]. Therefore, some of the fish farmers went to Canada. They said, 'We want bigger fish farms; we can do anything; we can do as we like.'
> —*Norwegian Parliamentarian. Transcript of the House of Commons Standing Committee on the Environment 12 September 1990*

The Norwegians by this time had spent unprecedented resources establishing an international export market for its farmed Atlantic salmon. It was not about to undermine that success by introducing a novel (in the eyes of their major markets at that time) product in the form of farmed Pacific salmon. The transition in British Columbia's aquaculture industry from Pacific salmon to Atlantic salmon was swift (Figure 5.1).

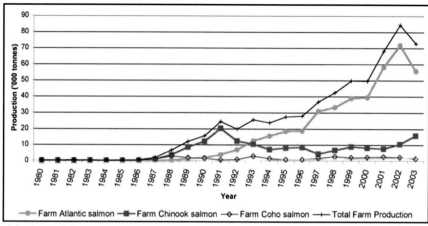

Source: Fisheries and Oceans Canada

Figure 5.1. Farm salmon production in BC by species 1980–2003. The first record of commercial salmon aquaculture in BC was in 1980 using coho salmon

Currently (as of 2006) Atlantic salmon comprises 76% of farmed salmon in BC (chinook salmon and some coho comprising the balance, both Pacific salmon species). Not only did the species change with the influx of Norwegian salmon farms, so did production volume. In 1988, during the early days of the industrialization process, BC farmers produced only 80,000 kilograms of Atlantic salmon; in 2002, with the industry in full blush, production exceeded 71,600 metric tonnes, representing an interim production increase of 895,000 times (Table 5.1). Today BC is the fourth largest producer of Atlantic salmon in the world, behind Chile, Norway, and Scotland respectively. While BC's industry has shown impressive growth over the past two decades, industry growth in competing countries was even greater, lead by Chile which in 1990 did not produce a single Atlantic salmon, but is currently the world leading Atlantic salmon grower, with 280,086 metric tonnes produced in 2003.

There is little doubt the tremendous growth of farm salmon production over this period played a significant role in the value erosion of wild product.

Table 5.1. BC wild salmon and farm Atlantic salmon production and value, contrasting changes from 1988 when Atlantic salmon aquaculture was initiated to the industry's peak performance period in 2002

Species	1988 Production (x1000 mt)	2002 Production (x1000 mt)	% Change	1988 Ex-vessel (CAD$ / kg)	2002 Ex-vessel (CAD$ / kg)	% Change
Chinook	5.92	1.69	-71	7.40	4.19	-43
Chum	30.30	12.35	-59	2.73	0.49	-82
Coho	7.08	0.47	-93	5.33	2.34	-56
Pink	32.22	8.61	-73	1.53	0.33	-78
Sockeye	11.94	10.15	-15	8.07	3.97	-51
	1988 Production (x1000 mt)	2002 Production (x1000 mt)	Change	1988 Farm Gate (CAD$ / kg)	2002 Farm Gate (CAD$ / kg)	% Change
Farm Atlantic	0.08	71,600	+895,000x	7.40	3.33	-55

Source: Fisheries and Oceans Canada

The massive global increase in farmed salmon production resulted in falling wholesale salmon prices, for both farmed *and* wild salmon. The immediate consumer benefit was clear; salmon had become affordable to a much larger proportion of the population. However, from the perspective of producers, particularly those in BC whose business plans had been developed in the heyday of salmon prices, the new reality of a globalized salmon commodity market has represented a significant challenge to fiscal viability. From 1988 to 2002 the farm-gate value of BC farm salmon dropped 55%. In order to overcome this precipitous decline, producers in BC as elsewhere, were forced to adopt ever-increasing economies of scale, the strategy being that in order to preserve a stable profit margin in a market of diminishing unit value, one must produce more units and do so as cheaply as possible. Production in 2002 increased 895,000 times over that of 1988. While the unit value of farm salmon had declined dramatically (55%); this was more than compensated for by the six orders of magnitude production increase. BC is by no means unique. The same scenario has been played out in all salmon farming nations, with each company in competition with all others to produce the most salmon for the least cost. The competitive profile of the industry is complicated by the fact that the vast majority of global production is generated by relatively few multinational companies, each operating in many countries and so pitting one region against another. It is this economic reality from which most issues currently associated with industrial aquaculture arise; the globalization of salmon has transformed this fish from a seasonal, high value delicacy to a low value commodity available year-round that comes with significant environmental—and thus public relations implications.

DISCUSSION

What are the implications of this situation for coastal communities and their food security? The high profile issue of the predominance of sea lice on salmon farms and the resultant increased infection of juvenile wild salmon is an example of how the globalization of salmon production can generate serious localized environmental impacts. In BC the term "sea lice" refers to either of two ectoparasitic crustaceans, *Lepeophtheirus salmonis* and *Caligus clemensi*. Both species naturally occur in coastal BC waters and thus are natural components of the marine ecosystem. Both are known to infect wild and farmed salmon in BC. The former is a salmon specialist and the latter a generalist, known to infect numerous non-salmonid fish species as well as salmon. The lifecycle of both types of sea lice is generally the same: fertilized eggs (up to many hundred per female) are attached to the female who is in turn attached to an adult salmon in the ocean. In time the eggs hatch into free-swimming nauplii larvae which in time grow and advance through copepodid, chalimus, and finally adult stages. Relatively early in this progression the young sea louse must find a host salmon (or perhaps one of a short list of other species in the case of *C. clemensi*). Given that the louse is free floating in the Pacific Ocean, the chances of literally running into a prospective host (not to mention successfully attaching) are extraordinarily slim. Therefore it is no surprise that natural sea lice abundances tend to be quite low. This scenario changes dramatically when salmon farms enter the equation. Salmon farms, by virtue of being home to as many as 1.5 million salmon in very small area, provide a high chance of success for any nauplii lucky enough to float through. Eventually such nauplii attached to their host mature and reproduce nauplii of their own. However, now the chances of successfully finding a host are very, very good—resulting in more farm salmon infections—and the cycle repeats itself until, if left unchecked, epidemic conditions rapidly set in. While this is clearly bad news for the farm salmon, how does this situation affect wild salmon?

We start by asking where are salmon farms situated? They are not randomly dispersed along the coast, but are placed in areas that, among other things, provide shelter from the open ocean (i.e. storms etc.). For this reason the head of bays and fjords are particularly attractive; but this is where the outlets of salmon bearing rivers are also found. In the spring and early summer when young wild salmon smolts migrate from their home rivers to the ocean they do so in the absence of adult salmon which are feeding offshore. Thus, under natural conditions, the absence of hosts in the inshore areas keeps juvenile lice abundance very low, allowing juvenile salmon to pass through with little impact. However, the situation can change dramatically if a salmon farm is located on the smolts' migration path. Research in the Broughton Archipelago has shown that an average farm can amplify local sea lice abundance 33,000 times over ambient levels, resulting in an infection rate of over seventy times what would normally occur. Further, this highly magnified farm "lice footprint" was detectable thirty kilometres from the salmon farm under investigation (Krkošek et al. 2005), considerably beyond the 'organic footprint' (Parrish et al., Chapter 3, this volume). The lice are spread even further as infected juvenile salmon continue on their migration out to sea. Most will succumb to their infection. However they may cover considerable distances before they do and all the while the once juvenile sea lice attached to and feeding on them grow and mature into reproductive adults. By the time these sea lice are ready to reproduce they may have traveled twenty or thirty kilometres from their originating farm. In the interim, new juvenile salmon, having not been exposed to a salmon farm, have joined the migrating school from other rivers along the migration corridor. But the now mature sea lice, having been picked up as nauplii thirty kilometres away, are now producing their own eggs and nauplii, and those young smolts recently joining the migration become infected. Krkošek *et al.* (1995) documented this phenomenon and witnessed how, through this process, a single farm infected otherwise unexposed salmon smolts more than thirty kilometres distant. These fish, in turn, extended the original farm's sea lice footprint to over seventy kilometres (the limit of the research survey). Dynamics of this nature are widely accepted as being responsible for spectacular collapses of Broughton Archipelago pink and chum salmon runs (Krkošek et al. 2005; Morton et al. 2004, 2005).

These data are no surprise to any one familiar with (1) fundamental epidemiology, or (2) the history of salmon aquaculture. It is very well understood that clustering hosts into exceptionally high densities and exposing them to contagion invites epidemics. Even if this theoretical expectation is not compelling, we need only look at the experiences of other jurisdictions that have gone before. Indeed, a very similar series of events occurring in Norway previously, and referred to above, precipitated the rapid expansion of salmon aquaculture in BC. Thus, we should not be surprised to see history repeating itself, particularly given that many of the founders of BC's current industry are those who resisted Norwegian government imposed reform in the aftermath of the *Gyrodactylus* incidents there.

Many have asked: "If these events were predictable, then why were they not avoided, or at least remedied following identification?" On the surface the answer appears complex and includes, among others, multi-scale ecological dynamics and aggressive political agendas. However, the actual underlying explanation is much more simple—sea lice epidemics (along with other environmental issues around salmon net pens) are in fact fated to occur and reoccur as a result of the globalization of the salmon industry.

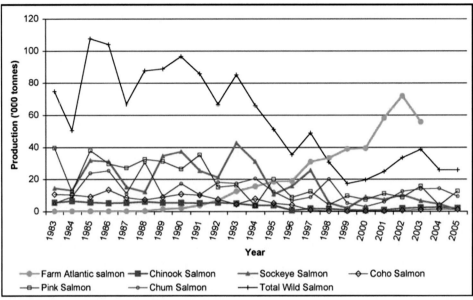

Source: Fisheries and Oceans Canada

Figure 5.2. Production of the BC commercial salmon capture fishery 1983–2005 by species plotted with production of BC farm Atlantic salmon. Data available to 2003 only.

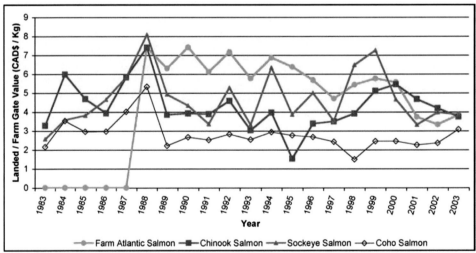

Source: Fisheries and Oceans Canada

Figure 5.3. Landed value of the top three most valuable commercially caught species plotted with farm gate value of farm Atlantic salmon 1983–2003

The most desired salmon farming areas on the BC coast are in waters where the adjacent land is wilderness (indeed this is typical of much of the coast). This is key, because it means there is no road access proximate to farms; all materials coming and going (feed, crew, product, waste, equipment, etc.) are conveyed by marine transport. This is exceptionally expensive relative to regular truck transport—which, thanks to the developed coastlines of Norway, Scotland, much of Chile, etc., is used to much greater degree in these regions. The reliance of the BC industry on marine transport rather than trucks represents a significant increase in production costs compared to those incurred by the international competitors, but BC producers cannot pass this cost on to consumers without losing market share. Therefore, this increases the nominal cost of farming salmon in BC. The rapid escalation of global farm salmon production has resulted in a worldwide glut of supply, ever diminishing the product value for BC producers (Figures 5.2–5.4) and resulting in a wide scale transition of salmon from high value seasonal delicacy to low value common commodity. The new reality of the global salmon commodity market tightly constrains options for adding product value in BC, leaving the only hope for maintaining economic solvency in adopting greater economies of scale and cutting production costs.

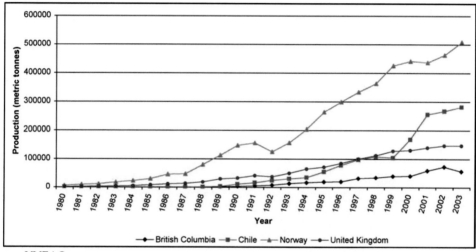

Source: UNFAO

Figure 5.4. Global production leaders of farm Atlantic salmon 1980–2003

The obvious first step in fighting sea lice epidemics in wild salmon is to adopt strict tenure sighting protocols to minimize cumulative farm effects including sea lice on wild stocks. In other words, farms should be spatially separated—as adopted in Norway following the *Gyrodactylus* outbreaks there. The incremental cost to Norwegian producers was modest. In BC, however, the salmon farm industry has reacted in exactly the opposite fashion. As value of farmed salmon drops, the need to cluster farms more and more tightly increases so as to minimize travel (i.e. cost) between tenures during the course of operations. This has resulted in the majority of BC salmon farm sites being highly clustered, with each cluster being located as close as possible to a town serving as a transportation hub. Major clusters are currently situated around Vancouver Island adjacent to Port Hardy, Port McNeil, Campbell River, and Tofino (Figure 5.5). Proposed industry expansion north will see Prince Rupert on the mainland added to the list.

The economic reality constraining the BC salmon farm industry is undeniable. The obvious solution to increased sea lice infestations would be to reduce farm densities, to move farms away from wild salmon smolt migration corridors and/or to fallow farms during the spring out-migration. Taking these actions, however, would not be economical.

This clash between environmental responsibility and economic viability represents a general pattern that is seen repeatedly across the array of issues associated with industrial scale salmon aquaculture. For instance, as value of farmed salmon erodes, the nominal *per capita* cost of an Atlantic salmon escapee diminishes too. Incentive for costly net maintenance declines as leaky net pens become more cost effective—revenue lost through escapes becomes less than the labour and material costs consumed to retain them.

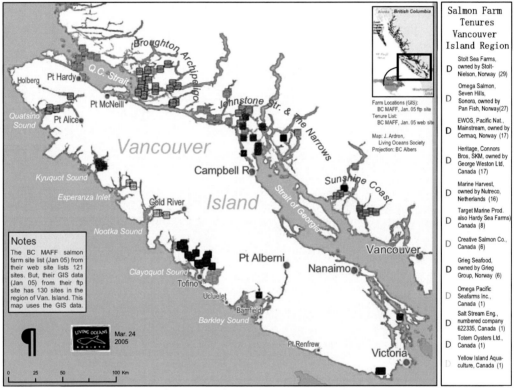

Source: Living Oceans Society <*http://www.livingoceans.org/index.shtml*>

Figure 5.5. Salmon farm tenure locations in BC. Note clustering of farms near transportation access points around Vancouver Island

Social dynamics are affected by product value erosion too. Labour is the second largest operating expenditure for salmon farms after feed. Falling salmon prices necessitate greater automation of the production system to trim costs. From 1997 to 2003 BC farmed salmon production increased approximately 100%, but on a per-unit production basis, wages, and salaries dropped 29% across the BC aquaculture sector (Cox 2005). Diminishing returns to coastal residents removes freedom to innovate and explore alternative production models leaving only intensifying industrialization as an option. In other terms, food sovereignty is forfeited and options are lost leaving communities with little choice but further investment in an industrial model with a bleak future.

Value erosion as a consequence of farm over production also sets in motion an analogous depensatory process in the salmon capture fishery. From 1988 to 2002 the production of every salmon species in the BC capture sector declined between 15% and 71%. Contrary to the basic tenets of economic supply and demand, as the supply decreased so did its value, up to 82% as in the case of chum salmon (*Oncorhynchus keta*) (Table 5.1). A simple analysis of catch values suggests fishermen today must harvest at a rate of up to five times that of 1988 to achieve an equivalent earning power. This is a conservative estimate owing to the fact that the 1988 figures have not been adjusted for inflation. Thus as prices drop, fishing pressure must increase to preserve earning power. Increased fishing pressure leads to smaller stocks which leads to increasing pressure and the depensatory cycle is in motion—greatly amplified by the negative market effects of an over abundance of farmed salmon.

On the surface, a multinational salmon farming conglomerate and an Amazonian rancher do not have much in common. However both have little sovereignty in how they make their living. The contexts are dramatically different; the underlying driving processes are the same. Both are locked in a battle, not against competitors for market share or restrictive government policy, but against a kind of mass psychosis so pervasive as to be iconic of modern societies: the myth of cheap. "Cheap" salmon or "cheap" beef in the case of the Amazonian rancher is an abstraction. In fact, "cheap" does not exist. Costs are transferred at various points in the production chain, but they do not disappear. The most damaging aspect of the industrial salmon farming experience is the establishment of a consumer culture expecting that salmon should be available fresh year round for the same price as chicken. And while this can be done, it is accomplished through massive economies of scale that demand significant ecological and social subsidies. These subsidies are not reflected in the retail price, but are costs just the same. Consumption of clean oxygenated water, removal of organic, and other wastes by currents, assimilation of escaped fish and greatly amplified numbers of sea lice in the ecosystem are just the start of the real costs transferred to the environment and to society at large.

Salmon farmers, like everyone else, would like to see the need for these subsidies eliminated. However, the reality is that farmed salmon is now a low value global commodity and that BC, relative to Norway or Chile, is an expensive place to farm salmon. The options for this dilemma are to either stay the course and pay the escalating social and environmental costs, or to abandon the industrial production model in favour of producing lower volume / higher value niche products (e.g. organic or "green" salmon). Both federal and provincial governments are clearly in favour of staying the course. The strategy seems to rest in the faulty belief that an as yet unattained "super-scale" of production will lead to sustainability.

Currently, as of 2006, we are weathering another major round of salmon farm industry expansion, extending north along the coast to Prince Rupert, in an effort to counter and compete with the rampant growth in Chile's farm salmon sector, which is showing no signs of slowing. The value of both farmed and wild salmon will undoubtedly continue on their conjoined journey of devaluing and decline, wreaking havoc enroute. The time for innovative development appears to be behind us now and those in the BC salmon sector—both farmed and wild—are little more the spectators, robbed of their sovereignty by shortsighted government policy driven by unrealistic consumer expectation. Ultimately, the loss of food sovereignty will translate to loss of food security, and this needs to be recognized by all those responsible for the salmon.

CONCLUSION

Three key points emerge from this analysis:

(i) Ecological issues that so often define the salmon farming controversy in BC and elsewhere, are in fact physical manifestations of underlying socioeconomic drivers. Working to solve the ecological problems while ignoring underlying drivers amounts to treating the symptoms while ignoring the disease. Efforts to conserve profitability in a globalized, commodity-based market (e.g. by increasing economy of scale) have resulted in simply transferring costs of production to associated ecological and social systems.

(ii) The fates of both BC salmon farmers and fishers are not in their own hands but are instead tied to global salmon market trends—both groups have seen their sovereignty eroded to the point of losing all control. Each, in their own way, is locked into reactionary strategies aimed at surviving the next round of incremental farm production increases without being able to plan for long-term sustainability.

(iii) Salmon farms in BC are forced to operate in what can be considered "hostile" territory. The physical realities of the BC coast impose production costs on BC operators that are not borne by their international competitors. Further, the social/cultural climate of coastal BC is increasingly unfriendly, resulting in further escalations in the cost of doing business here—again a burden not borne by foreign competitors. These factors, together with the seemingly limitless potential for expansion in other jurisdictions, particularly Chile, result in the future of industrial salmon farming in BC looking very bleak. Furthermore, industrial salmon farming is weakening wild salmon production due, for example, to the degrading effects of escalating sea lice populations and escapement of Atlantic farmed salmon into Pacific salmon habitat. If these effects are not remedied soon, both salmon sectors are threatened. For the First Nations and others who rely on wild salmon for cultural reasons as well as nutritional, the prospects look even more grim.

After consideration, one sees that sovereignty within the BC salmon industry, from perspective of both fish farmers and fishers of wild salmon, has been dramatically compromised and in all likelihood is beyond recovery so long as current market trends continue and consumers fail to discriminate between the full costs of industrial and sustainable salmon products.

In: Resetting the Kitchen Table
Editors: C. C. Parrish et al., pp. 87-98

ISBN 1-60021-236-0
© 2008 Nova Science Publishers, Inc.

Chapter 6

NEW FOODS, LOST FOODS: LOCAL KNOWLEDGE ABOUT CHANGING RESOURCES IN COASTAL LABRADOR

John C. Kennedy

Department of Anthropology, Memorial University of Newfoundland, Canada

ABSTRACT

Anthropological research during the autumn of 2000 restudied communities in two regions of coastal Labrador, Canada, where the author has worked for over thirty years. The Labrador people interviewed reported numerous changes in the land and sea resource base that forms the foundation of their resource-harvesting lifestyle. Resources such as salmon and cod that people once consumed or traded are now rare, while others, such as snow crab and shrimp, have acquired new importance. Labrador people also report that the abundance and behavior of other natural resources—including capelin, harp seals, and wolves—has changed, and they do not know why. After a brief overview of the regions and communities studied, the chapter presents some examples of local knowledge about changes in land and sea resources, and local explanations for such changes. The chapter concludes by briefly discussing some possible scientific explanations for the changes the Labrador people have perceived. In particular, a regime shift in the waters of the Northwest Atlantic is suggested as a possible cause for some of the identified changes.

INTRODUCTION

This chapter presents data gathered on the coast of Labrador (see Figure 6.1) during the autumn of 2000 as part of the *Coasts Under Stress* (*CUS*) research project.[1] I present a small

---

[1] I thank the Social Sciences and Humanities Research Council of Canada (SSHRC) and the Natural Science and Engineering Research Council of Canada (NSERC) who have provided the major funds for the *Coasts Under Stress* (*CUS*) Project through the SSHRC Major Collaborative Research Initiatives program. Memorial University, the host university, provided some funding. Unless noted, all opinions and interpretations are those of the author, and should not be considered representative of others associated with the *CUS* project. I thank

sample of local opinions about what people consider to be a rapidly changing environment. Many land and sea resources that have long been used by people for food and for trade have either disappeared or become scarce, according to their observations, while other, formerly unused resources, have acquired new importance.

The best example of this is the demise of the northern cod fishery, commencing with the moratorium of 1992. People now see the vulnerability of other resources through their experience with cod. Thus, for people long accustomed to a level of predictability, notwithstanding the vagaries of seasonal change, one stress they now experience is uncertainty about the future. Uncertainty is not unique to the people of coastal Labrador, but the isolation and resource dependency of the coast's small communities magnifies the impact of insecurity, clouding individual and collective decisions about how to plan for the future. The Labrador people I interviewed grapple with the question of why environmental change is occurring, and because people often speculate about causation when describing specific changes, I have grouped descriptions of change and cause together, separating them only by the categories of land and sea resources. Finally, I conclude with a brief overview of some of the scientific interpretations of why the resources of Labrador's terrestrial and marine environments may be changing.

In my return to Labrador in 2000 I investigated how coastal communities had changed over the thirty years since I began my research there, in the northern Labrador community of Makkovik, in 1971. I also retraced my footsteps in several communities within southeastern Labrador, so it is best to begin with a brief overview of these two regions and of the communities I visited.

Southeastern Labrador

Southeastern Labrador includes the region between Lodge Bay and Cartwright (see Figure 6.1). Historically, fisher families of the region practised a migratory or transhumant settlement pattern and seasonal economy once widespread throughout Newfoundland and Labrador (Smith 1987). As in northern Labrador, European settlers, males recruited to work in Labrador's fishery or fur trading businesses, took Inuit or part-Inuit women as spouses, explaining why people within the region now call themselves Métis, that is, of mixed Inuit-European ancestry (See Kennedy 1997). Families wintered in sheltered coves and bays, near sources of food, forests, and water, and moved each spring to the headlands, islands, and capes on the outer coast, where they harvested harp seals, migratory waterfowl, salmon, cod, and herring, more or less in that seasonal order. These marine resources were purchased by local merchants and also constituted most of the daily diet consumed by fisher families. Consequently, until the end of the cod fishery in 1992, southeastern Labrador peoples were indeed, *of* the bays and headlands (Kennedy 1995).

my 111 Labrador informants (60 males, 51 females), who must remain anonymous, and who patiently gave of their time and knowledge. I also thank my MUCEP student Janice Brake; Mr. Wayne Barney and Terrance McCarthy of the Newfoundland and Labrador government; Larry Yetman, Earl Dawe, Dave Taylor, Becky Sjare, Gary Stenson, Regina Anthony, and Judy Dwyer of the federal Department of Fisheries and Oceans; and Liz Smith of the Professional Fish Harvesters Certification Board. Finally, in thanking Dr. Bill Montevecchi and the volume's editors for their useful comments on earlier drafts, I accept responsibility for all errors of fact or interpretation.

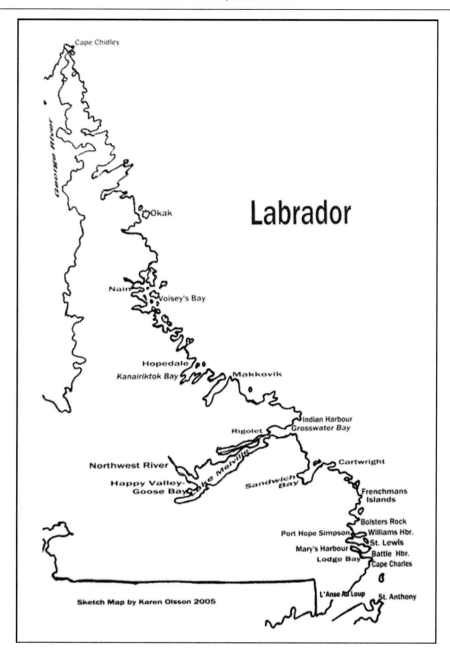

Figure 6.1. Coastal Labrador

However, during the late nineteenth and early twentieth century, changes (both natural and cultural) increasingly threatened this migratory and highly seasonal way of life. For example, during the 1870s, cold waters ravaged the fishery, led to the sales of local mercantile establishments, and caused some out-migration (Kennedy 1995, 99–100). The 1918–19 global pandemic of Spanish Influenza following the First World War killed around seventy people in Sandwich Bay, requiring an orphanage that was eventually located at Cartwright. Fish prices plummeted at the onset of the Great Depression in 1929, reducing Newfoundland's Labrador fishery, and causing the three-decade-old Grenfell medical mission to move "outside" hospitals such as Battle Harbour and Indian Harbour "inside" to sheltered bays. After Newfoundland and Labrador joined the Canadian federation in 1949, a decisive policy of population concentration and resettlement reduced the number of "inside" winter communities. Educational policies regarding school attendance soon reduced time spent at outside fishing stations. Yet, not withstanding these and other changes, southeastern Labrador's migratory settlement pattern and seasonal economy survived until the end of the cod fishery in 1992. Then all changed.

Northern Labrador

Northern Labrador comprises the coast from Rigolet north to Cape Chidley (see Figure 6.1). Historically, Inuit (Eskimos) were colonized by Moravian missionaries in 1771 when that church established the first in an eventual chain of mission stations serving Inuit and later European settlers along the northern coast. Permanent European settlement of much of coastal Labrador began in the late eighteenth and early nineteenth century. In both southeastern and northern Labrador, many Europeans took Inuit spouses, although such unions were more common and certainly better known in northern Labrador. Although Moravian missionaries initially opposed European settlers entering what they considered their territory, fearing the corruptive influence of Europeans on the Inuit, after the 1850s the Moravians accepted settler converts, although the church always considered settler and Inuit parishioners distinct. Like their counterparts in southeastern Labrador, northern Labrador Inuit, and settlers moved seasonally to harvest seals, various species of fish and birds, caribou, and other resources, both for domestic consumption and for trade at the Moravian and later, at the Hudson Bay Company and government stores. In the fall of 1918, someone aboard the Moravian trading ship unintentionally carried Spanish Influenza to northern Labrador, where a lack of immunity among Inuit proved lethal. Between November and January, influenza killed approximately one-third of all Labrador Inuit, causing the missionaries to close the community of Okak, and to recruit Inuit from other communities in order to repopulate communities emptied by death (Tanner 1944). In 1926, the Moravians leased their trading posts to the Hudson's Bay Company (HBC). Bullish global demands for furs led the HBC to press Inuit for fox skins, and this required trapping families to abandon their more generalized economy and diet, and to purchase European foods; Inuit health deteriorated (Kleivan 1966, 128–36; Jenness 1965, 60–61; Tanner 1944). Confederation in 1949 improved social services but at cost: the new provincial government closed two northern Inuit communities and relocated Inuit southward, to become "strangers" in settler or Inuit villages (Brice-Bennett 1994).

Communities Visited in 2000

The 2000 research in southeastern Labrador occurred in four of the region's eleven permanent communities: Lodge Bay, Mary's Harbour, St. Lewis, Port Hope Simpson, and in the northern Labrador community of Makkovik. Each community has a unique history.

What is now Lodge Bay was used as a winter settlement as early as the French Regime, between 1702 and 1763. However, the roots of today's community date from the early nineteenth century, when Lodge Bay was the winter community of people who fished each summer at Cape Charles, on the headlands around twenty kilometres to the east (Kennedy 1996).

Mary's Harbour, just north of Lodge Bay, was founded in 1930 by the Grenfell Mission as an "inside" hospital community to replace the Mission's first hospital at Battle Harbour, which burned the previous year. As at Cartwright, North West River, and St. Anthony, Grenfell Mission leaders such as Dr. Harry Paddon intended Mary's Harbour to have large gardens, both as an alternative to the fishing and trapping economy and to raise nutritious vegetables to combat deficiency diseases, such as beriberi, scurvy, and rickets (Sniffen 1923, 110; see also Solberg et al., Chapter 11 and Kealey, Chapter 12, this volume). Mary's Harbour grew slowly until the provincial resettlement program of the 1960s, when families from Battle Harbour, Indian Harbour and Matties Cove moved to winter there. A salt fish processing plant was built in Mary's Harbour in the 1970s and converted a decade later to process snow crab, for export.

St. Lewis (or Fox Harbour) first appears on an early sixteenth-century map as Ilha de Frey Luis (Gosling 1910, 61–62) and was described by nineteenth century visitors such as DeBoilieu (1861/1969) as an "Eskimo [Inuit] village." St. Lewis mushroomed in 1955, however, when the United States Air Force began construction of a Gap Filler radar site overlooking the community, attracting people who formerly wintered in St. Lewis Bay. A succession of cod processing plants preceded the present crab plant, operated by Coastal Labrador Fisheries.

Port Hope Simpson is Labrador's second (to Mud Lake, in central Labrador) modern company town, established by the Labrador Development Company in 1934 to cut pit props to reinforce subterranean Welsh coal-mines (Kennedy 1995, 158–72). Work at Port Hope Simpson attracted Labrador people from nearby bays and workers from Newfoundland. Following confederation and the collapse of the company, the Bowaters Company cut and shipped round logs to Newfoundland and England between 1962 and 1969. Small scale, local woods contractors followed. During the decades immediately preceding the cod moratorium, people cut wood during winter, and fished at outside communities during summer (Southard 1982).

Finally, the northern Labrador community of Makkovik was settled by a Norwegian man and his Labrador wife in the 1850s. The Moravians established a church there in 1896 to serve the predominantly settler population south of Hopedale. In the 1950s, northern Inuit were relocated to Makkovik, creating the ethnically divided community observed by Ben-Dor (1966) and Kennedy (1982).

LOCAL KNOWLEDGE ABOUT THE CHANGING TERRESTRIAL ENVIRONMENT

Many people observed that while normal amounts of snow fell during the winter of 1999–2000, there had been less snow during the previous five winters. An elderly Lodge Bay woman opined, as did others, "There is not as much frost or snow as we used to have." A Port Hope Simpson fisher recalled that the last three summers (1998–2000) have been warmer than usual, with more southwest winds. The summer of 1999 was both warm and dry, with reduced water levels in local rivers that prevented salmon and Arctic char from entering their natal rivers. A Port Hope Simpson woodsman attributed what he believes is a faster growth rate in trees to warmer temperatures. Generally then, people claim that the climate has moderated in recent years.

People had little to say about changes in ground vegetation, or indeed about the small animals forming the middle of Labrador's terrestrial ecological pyramid (Elton 1942). A Port Hope Simpson forester spoke about blights common to second-generation trees, but his concern was more with the long-term consequences of Bowaters' forestry practices than with natural changes. In his words, "they [Bowaters] selectively cut (high grading). If they cut a tree, and it fell into another grove, it damaged the wood we're now cutting. Damages like curls, twists, warp all lowers the grade" [producing what locals call "depression wood"].

People commented on the cyclic availability of some species, and on shortages of others. A Port Hope Simpson trapper spoke about martens (*Martes americana*), recalling, "around 1989–90, martens were thick, then scarcer, now thicker again." The same man described how forest fires in the 1980s reduced numbers of lynx (*Lynx canadensis*), which he claimed are now coming back. Numerous the past two years, rabbits (varying hare, *Lepus americanus*), a very important local source of meat, appeared rare during the autumn of 2000. A Lodge Bay man explained that they "are not down yet, they're now up in tuckamores [wind-stunted spruce, on high ground] and not down [in the valleys] yet."

Another reason for the scarcity of rabbits in the autumn of 2000 could be the greater numbers of foxes (*Vulpes fulva*) and wolves (*Canis lupus*) recently seen in or near communities, especially in summer and autumn. During the years immediately preceding my research, encounters with foxes and wolves had been common in many Labrador communities. Indeed, fear of wolves caused the northern Labrador school at Nain to be closed a few days in autumn 2000. Locals describe these animals as "brazen" (bold) and curious. A Mary's Harbour man jogging along the road to Lodge Bay was followed by a fox that stopped every time the man stopped, and was oblivious to the rocks he threw at it. While no one was sure why foxes and wolves abound, some people offered explanations. One Lodge Bay man observed that the foxes "must be hungry," yet there is no evidence of household garbage being eaten. Formerly there were wolves once in a while but, he said, "we've never known foxes in the community." Others claimed that the construction of the Trans Labrador Highway between Red Bay and Cartwright (see Gibson et al., submitted) frightened foxes and wolves eastward of the new road. Some blamed garbage at highway construction lunch sites, citing cases where workers hand-fed foxes from lunch pails. In the summer of 2000, a Lodge Bay man saw five wolves on the new highway, about twenty-five kilometres south of Lodge Bay. He also described a circuit wolves appear to make, traveling from Simms Bay (south of Lodge Bay), to Lodge Bay, around Long Pond to Mary's Harbour and back. The placement of Simms Bay on this circuit is intriguing, since locals explained that since around 1990, harp seals (*Pagophilus groenlandicus*) frequent the frozen "landwash" (sea shore) ice of Simms Bay (south of Cape Charles) in November and December, something older people had never seen. A Lodge Bay man said that in Lewis' Bay, harp seals trapped in ice near the land, have been hauled into the woods, in his words, "wolves taken 'em from ice near open water holes."

Of course wolves are usually associated with caribou (*Rangifer tarandus*) herds, which mysteriously disappeared from the southeastern Labrador interior around 1953, the same year that the Newfoundland government introduced moose (*Alces alces*) from Newfoundland to St. Lewis Inlet as a food source. Quite possibly, southeastern Labrador wolves are thinning out weak or old moose, although most people claim that the local moose population has increased since 1995, when locals began crossing the Strait to hunt moose in autumn on Newfoundland's Northern Peninsula (Area 40). Both Newfoundland moose, and Labrador caribou, the latter hunted from the Goose Bay area, appear on southeastern Labrador dinner tables.

A Lodge Bay man spoke of "rumors" of caribou inland from L'Anse au Loup, a Strait of Belle Isle community. Similarly, a Fox Harbour man claimed that within the past ten years, a herd of caribou was seen around sixty-five kilometres inside of Riverhead (St. Lewis Inlet), that caribou tracks were also seen at Caplin Bay three to four years ago, and that caribou tracks were observed at Hills Harbour (south of Punch Bowl) and at Bolster's Rock in the spring of 2000.

LOCAL KNOWLEDGE ABOUT THE CHANGING MARINE ENVIRONMENT

Labrador people claim that some seabird species are now extremely scarce compared with previously. In southeastern Labrador people characterize the eider duck (*Somateria mollissma*) population as "never being lower," perhaps because of excessive hunting (see Montevecchi et al., Chapter 7, this volume). Unlike the older wooden speedboats formerly used to hunt seabirds on the ocean, today's fiberglass boats can be operated safely in ice-filled sea waters, increasing hunting pressure on eiders, turrs (or Common Murre—*Uria aalge*), and other seabirds. A Lodge Bay man claimed that "ducks don't taste so good, there's an odour when cooking. Not the same, now, drove [chased by boats] too much."

I could not enter a house without hearing reports of the unusual seasonal behavior of caplin (or capelin, *Mallotus villosus*), the diminutive fish that is a foundation of the marine ecosystem. All locals describe recently observed caplin as smaller than normal, as remaining inshore until late autumn, and as occasionally spawning in deep water. A retired Lodge Bay fisher has recently seen caplin in the intestinal tract ("puddick") of turrs in October and November, something not formerly seen. Another retired Lodge Bay fisher argued that caplin are overfished, and attributed what locals regard as their present aberrant distributions to the fact that there are no cod (*Gadus morhus*) to drive them ashore (see Eythorsson 1998, for a similar, albeit Norwegian, example of this). The man explained that historically, inshore cod migrations occurred in two phases: the first schools of cod were eating shrimp in deep water, while the second school of cod were chasing caplin toward land. He said that when he first saw cod with shrimp in them he wondered what the shrimp were.

Labrador people appear to recognize two kinds of caplin, bay caplin, and outside caplin. "bay" or "winter" caplin is described as being smelt-sized, and resident in particular bays. As a Lodge Bay man put it: "there's caplin that stays all winter long in the bays. In Lewis' Bay in May month, you can catch trout with caplin in them. And when father sealed in December, they'd be small caplin, sometimes called "white fish" in the seals. Winter caplin has no ridge on them, just the same as a spawned caplin, no ridge on his sides." A Makkovik fisher added details. Bay caplin, he explained, are smaller, brown-light on their back while outside caplin, are larger, dark blue-green, and around twenty-five to thirty centimetres in length. Outside caplin spawn on beaches, such as they did at Caribou Point (near Battle Harbour) in July

2000, often during an easterly wind. Bay caplin spawn in shoal (shallow) water in bays, and their spawn can be seen at the low tide mark. Although one of the Makkovik Inuit I interviewed maintained that there have always been bay and outside caplin, the Inuit language (Inuktitut) does not have separate names for them; both are *Kuliligak*. Another Makkovik man described both types of caplin, claiming that while bay caplin are always around, outside caplin are not common, in his words: "I haven't seen one in ten to fifteen years."

In Lodge Bay, Mary's Harbour, and Port Hope Simpson I heard many accounts of the caplin that appeared near the dock in Mary's Harbour (at Gin Cove) in November 1999. Most remember these caplin as small and opinions vary on whether they were spawning. A Lodge Bay man recalled with laughter how a Mary's Harbour man had claimed the "caplin spawned, the beach is full of spawn, all cock [male] caplin but the beach is full of spawn!" A Port Hope Simpson man, who ate some of these caplin, claimed they were both sexes and small, approximately two-thirds the size of "normal" [presumably "outside"] caplin. A Lodge Bay man dated the Gin Cove incident to 12–19 October 1999, recalling that the caplin were not spawning, and were mostly males. As in 1999, caplin were at Nimrod Tickle (just inside of Battle Harbour) in mid-October 2000, attracting seals and whales.

All along the Labrador coast, Atlantic cod has been rare, stunted in growth, or absent from customary locations. During recent years, what Labrador people know about cod comes primarily through the Sentinel, Index, and food fisheries. With infrequent exceptions, everyone describes the cod caught by these fisheries as small (the size of "rounders," that is, whole salted cod roughly thirty to forty-five centimetres). The exceptions I heard include one catch of cod measuring fifty to fifty-five centimetres at William's Harbour in summer 2000 and a 1999 catch of eleven cod at Old Man Shoal on the north side of Makkovik Bay. All these eleven fish measured between forty and fifty-three centimetres. Although these exceptions are encouraging, the rule is more sobering. According to the DFO Fishery Officer in Makkovik, only twelve cod were taken in the 2000 food fisheries occurring between Makkovik and Hopedale. The disappearance of cod is an event of monumental cultural and economic importance. Salted and frozen cod formed a vital part of the diet, a staple now replaced with purchased foods. One former Cape Charles fisher complained that now "a fella has to shop to buy a bit of fish but the quality is poor." This new dependence on commercially produced food fuels local concerns about nutrition. In the words of a Lodge Bay woman, there is "more talk of cholesterol since the [cod] moratorium, especially among the older people."

Locals describe Atlantic salmon (*Salmo solar*) as plentiful in local rivers during recent years, although one wonders whether such claims are rhetorical, vindicating local fishing in nearby rivers. In any event, local exposure to salmon is now twofold. First, since the closure of the cod, salmon and Arctic char (*Salvelinus alpinus*) fisheries, the associated abandonment of outside fishing stations, and permanent residence at "winter" communities that are often located near salmon rivers, many former commercial salmon fishers now use fly rods to angle for salmon in their natal rivers. This change is significant because of the frequently heard local insinuation that outside "sport" anglers only "torment" salmon. Today, Labrador's new salmon anglers insist that their sole aim is food, not sport. Secondly, salmon food fisheries, and some by-catch of salmon during the trout fishery provide people with a different (debatably more accurate) opportunity to monitor salmon. In one Port Hope Simpson fishers' words: "there is almost as much salmon caught now as when we had a commercial fishery." A Makkovik man netting harp seals in Makkovik Bay, from mid-November to early

December 2000 unintentionally caught eighteen large (the smallest was 4.5 kilograms) salmon in large mesh (twenty to twenty-six centimetre) seal nets. This accidental salmon by-catch, along with local claims that salmon are plentiful are a promising counterpoint to the predominant verdict that Atlantic Salmon are endangered. But again, these optimistic assessments from Labrador people may be a result of their year round residence near the natal rivers where salmon must enter, as opposed to observing or catching salmon further out in the bays or near outside headlands.

Labrador people are learning about species such as sea urchins (*Stongylocentrotus purpuratus*), whelks ("wrinkles" *Buccinum undatum*), scallop (*Chlamys islandica)*, snow crab (*Chionoecetes opilio*), and northern shrimp (*Pandalus borealis*), which constitute the current fishery (see Kennedy, forthcoming). Local knowledge about these "new species" is accumulating but limited. Most fishing for snow crab occurs well over one hundred kilometres offshore, in deep water. On the basis of what fishers learn from pulling their crab pots, crab prefer soft, boggy ocean bottoms. Snow crab and shrimp grounds overlap, with the implication that crab netted where shrimp are fished are sometimes mangled, lowering their market value. The high cost of crab and shrimp means that neither commonly appears on local tables.

People from Lodge Bay to Makkovik claim that the populations of several species of sea mammals have increased, and that the range and health of several species appear to have changed. The most dramatic change involves harp seals, historically the most important of Labrador's three migratory species [i.e., harps, hooded (*Cystophora cristata*), and grey (*Halichoerus grypus*) seals]. Historically, harps began their migration northward to the Arctic or Greenland in May, following the birth of pups off southern Labrador (Mansfield 1967, 13). However, my informants consistently report that some harps no longer migrate, remaining instead along the Labrador coast.

Harp seals were seen at Indian Harbour (near Battle Harbour) and were "everywhere" around Battle Harbour in July 2000. They were also seen around Makkovik in summer 2000, in Big River (south of Makkovik) in late May and early June of 2000, and elsewhere. In the words of the DFO Fishery Officer in St. Lewis, "All summer long we see harp seals in the bay." Harps, especially bedlamers [adolescent harps], are also common during autumn. A Makkovik man described thin young bedlamers (which he called "bluebacks," a name sometimes applied to hooded seals [see Lien et al. 1985, 106]) in fall having no fur coat on them, and occasionally greenish polyps under their forelimbs. These bedlamers are usually described as possessing little body fat, and as chasing smelt, caplin, herring, and other bait. One Makkovik man linked caplin and seal distributions. In his words, "Anywhere there's caplin, seals will bide [remain]." In places like Kanairiktok and Voisey's Bay harp seals linger longer, eating bay caplin. On 27 October 2000 the waters around Cape Charles were alive with whales, harp seals, and sea birds. Two hunters shot three harp seals for food. The scant, partially digested food in their stomachs appeared either to be "white fish" (caplin) or, perhaps, smelt. On the other hand, a harp seal killed at Mary's Harbour around Christmas 1996, is reported to have had eaten 16 small codfish, each around thirty-eight centimetres long.

In early winter, as the bays freeze, harp seals are often trapped in new ice some distance from open water. A Port Hope Simpson man said that seals become " . . . frozen in bays, and have always gone into the woods, because it's dark, like water." A Makkovik man added that harps trapped in bays "take to the woods [because] they can't see well, and go toward the dark." Local explanations of why migratory harps remain longer in Labrador concentrate on "bait," that is, on the local availability of food sources holding the harps longer. This raises the question of why, if food sources are sufficient to hold harp seals longer in Labrador, the size of the few cod taken locally (as reported above) is so stunted and the harps so lean?

Locals believe that grey seals are more common than formerly, whereas ringed (*Pusa hispida*) and harbour (*Phoca vitulina*) seals, so important to the historic economy and diet, have declined. A Makkovik Inuk (an Inuit man) lamented the decline in ringed (jar) seals as follows: "we don't see jar seals, especially in the fall, but they can be seen in the spring. I don't know why. Jars were our main species we hunted. Now mainly harps [that is, harps are prevalent]. We used to see jars in the fall, now only on the ice. We miss seeing those jar seals. It was good to see them. You knew you had a meal. I think the jars are moving north of here. People are talking about a lot of jars north of Nain."

Some species of small whales (Lien et al. 1985, 73–93), notably the common or harbour porpoise ("herring jumpers" or *Phocoena phocoena*) and white-sided dolphin ("jumpers" *Lagenorhynchus acutus*) are said to be more common than in former years. The same applies to larger whales, such as the killer whale (orca, *Orcinus orca*). A retired Lodge Bay fisher told me that he saw his first killer whale in the 1980s and has seen many since. All agree that some whales, particularly humpbacks (*Megaptera novaeangliae*), are, as one Port Hope Simpson fisher put it, "thicker [more common] than when we fished (for cod)." Just as these species of whales have increased so they have extended their stay in local waters. As a Port Hope Simpson fisher put it, "Whales used to be here in June and July, now it's later in the fall."

SCIENTIFIC KNOWLEDGE ABOUT LABRADOR'S CHANGING TERRESTRIAL AND MARINE RESOURCES

My familiarity with scientific data explaining the kinds of recent changes locals report is limited, although I discussed the local knowledge summarized above with Mr. Wayne Barney, Species Management Coordinator with the Provincial Wildlife Division, as well as those listed in my acknowledgements. Regarding changes in the habits of terrestrial mammals, Mr. Barney explained that the scientific literature on the kinds of recent changes reported above is scant. However, there has been a documented southward radiation of the George River caribou herd since around 1985. Moreover, Mr. Barney corroborated informant's caribou signs at Caplin Bay, suggesting these small groups of caribou probably belong to the small Mealy Mountain caribou herd. Increased sightings of foxes may result from periodic crashes of interior populations of small rodents such as the northern red-backed vole (*Clethrionomys rutilus*), causing foxes to move toward the coast. Decreased fur prices have reduced trapping, and the situation is further complicated by an outbreak of rabies in 2000–01. On the other hand, Mr. Barney believes that the wolves, people report, may have initially followed the George River caribou herd southward and have remained in southeastern Labrador because of the area's small moose population. Occasional wolf predation on trapped harp seals, although possible, is apparently not scientifically documented.

Scientific knowledge explaining changes to marine resources may be more complete than that pertaining to land resources. Regarding the two kinds of caplin people describe, Templeman (1966, 106) includes the vernacular "whitefish" as a common name for caplin, notes sexual dimorphism, and describes deep water spawning, but does not mention two types of caplin. On the other hand, I am uncertain whether the "Avalon" and "Labrador" caplin stocks Nakashima (1992) discusses relate to the two kinds of caplin Labrador people identify.

The Labrador people I interviewed know a tremendous amount about their environment but were divided about how to explain the kinds of environmental changes summarized above, or had no explanation. Like all forms of knowledge, local knowledge is finite and changeable. Researchers need also realize that informants' statements about local knowledge should not be separated from the broader political discourse about environmental issues. More generally, environmental change is usually attributed to human causes, such as global warming or, in this case, over-fishing. But over-fishing may not explain why harp seals, caplin, or whales linger along the Labrador coast, prompting me to mention what scientists call a "regime shift" which has cooled Labrador Sea waters since the late 1960s.

The waters of the Labrador Sea (the Northwest Atlantic Ocean off Labrador) are stratified, containing three main layers. Data supporting a regime shift comes at least two sources. First, the work of fellow *CUS* researcher Dr. W.A. Montevecchi demonstrates how the ocean foods seabirds (e.g., gannets [*Sula bassanus*], guillemots [*Cepphus grylle*], kittiwakes [*Rissa tridactyla*], and others) eat reveal changes in fish distributions and populations. An empirical example of this methodology and perspective concludes that northwest Atlantic surface water temperatures cooled during the early 1990s, influencing the timing, movement patterns, and accessibility of pelagic fishes (Montevecchi and Myers 1996, 313). A second source also concludes that the waters of the Northwest Atlantic have cooled. Thus, long-term studies of hydrographic records suggest that since the late 1960s, slower convection during winter cooled, freshened, and deepened the intermediate Labrador Sea layer. The resulting cooler waters diffuse southward past Newfoundland and eastward toward Europe (Sy et al. 1997; Dickson 1997). Rapid spreading of these cool intermediate waters influences global climate but more locally, explains the increased incidence of polar or Arctic cod (*Boreogadus saida*) in inshore Labrador and northern Newfoundland waters (Lilly et al. 1994). Arctic cod eat Atlantic cod spawn and the caplin-sized Arctic cod are themselves eaten by harp seals, particularly in inshore waters (Lawson et al. 1998). Greater concentrations of Arctic cod may have convinced harp seals to linger in Labrador, yet this still fails to explain why the harps Labrador people kill or net in autumn are so lean. Even though the relationship (if any) between cooler waters and the failure of Atlantic cod to recover, and that between Arctic cod and changing caplin distribution is unclear, the timing of this cooling regime and changes in the Labrador marine ecosystem are intriguing and may ultimately explain some of the changes in the marine environment that locals have observed.

CONCLUSION

The local voices presented above conclude that Labrador's land and sea environments are changing, leading to uncertainty about the future of local resources people once took for granted. The collapse of Atlantic cod stocks is of course the best known of these changes but the Labrador people I interviewed identified a host of other changes, which has meant that some foods formerly eaten regularly are now rare or absent altogether from dinner tables. The end of the cod fishery and the crab fishery that replaced it also means that many former cod and salmon fishers (and their spouses) now work on crab assembly lines, producing crab for markets in the United States, Japan, and elsewhere (see Kennedy, forthcoming). This leaves little time for harvesting local resources but then again, many of these are now rare. The seasonal scheduling of time that I observed in the 1970s has been replaced by shift work at a crab or shrimp plant, at least during the summer duration of these fisheries. All this means that what people do, what they eat, and the meaning of their lives has changed, much like the environmental changes they report. Whether the environmental changes people report are the result of human activities, natural processes such as a regime shift, or some complex amalgam of cultural and natural causes cannot as yet be determined.

In: Resetting the Kitchen Table
Editors: C. C. Parrish et al., pp. 99-113
ISBN 1-60021-236-0
© 2008 Nova Science Publishers, Inc.

Chapter 7

HUNTING FOR SECURITY: CHANGES IN THE EXPLOITATION OF MARINE BIRDS IN NEWFOUNDLAND AND LABRADOR

W. A. Montevecchi[1], H. Chaffey[2] and C. Burke[2]

[1]Psychology, Biology, and Oceans Sciences,
Memorial University of Newfoundland, Canada
[2]Cognitive and Behavioural Ecology Program,
Memorial University of Newfoundland, Canada

ABSTRACT

North American wildlife exploitation, as exemplified in the seabird and seaduck hunts of Newfoundland and Labrador, was a basic means of food security in coastal communities. Patterns of need and exploitation changed radically since the arrival of Europeans who perceived abundant and inexhaustible wildlife populations. These perspectives often combined with adversarial approaches of securing livelihoods by "conquering" the wilderness and its aboriginal inhabitants. Unrestrained harvesting and notions of free public access were the antithesis of aristocratic land ownership in Europe that often denied people in need access to wildlife. The new North American ideals also ran counter to conservation initiatives such as hunting restrictions that were viewed as unacceptable government control.

Technological improvements in transportation and in fishing and hunting capabilities (e.g. longer ranging, faster vessels, automatic weapons) helped to secure food over larger spatial scales and to bolster larger economies. Improved technology also created breakpoints in wildlife exploitation that led to over-harvesting. Conservation legislation developed gradually, but lagged behind the decimations of many wildlife populations. The reality and often finality of overexploitation were realized slowly, though not usually heeded. In the case of marine birds, unsustainable cumulative mortality from hunting, fishing, and oil pollution eventually resulted in the implementation of comprehensive conservation laws and regulatory policies.

Through the twentieth century, hunting for food security shifted to essentially recreational forms of hunting. Yet many households in coastal communities still supplement (at times substantially) family provisions with wildlife. Overall, interest in hunting is waning in Newfoundland and Labrador (and North America), due in large part to out-migration from coastal communities.

Non-consumptive uses of wildlife are of much benefit to coastal communities through ecotourism. Hunters, fishers, ecotourism operators, and others are involved in *Coasts Under Stress* participatory research to protect wildlife populations, to preserve and enhance critical marine habitat and to help sustain coastal communities. Interdisciplinary environmental approaches and educational outreach are integrating natural and social sciences and local ecological knowledge (LEK) in developing effective conservation policy (Jenkins 2003). As the footprint of human activity expands, wildlife and environmental conservation is a vital life-sustaining strategy.

INTRODUCTION

Efforts to secure food and community survival are globally manifest in geographic associations between environmental resources and both ancient and current human settlements (Freuchen and Salomonsen 1958; Diamond 1997). For coastal communities, location and persistence have been determined by the availability, abundance, and predictability of edible marine animals and plants (McGhee and Tuck 1975; Harris 1990).

Owing to dependence on wildlife for survival, hunting methods and technologies were aimed at improving reliability, and efficiency, and ultimately on increasing food security. Improving technology restructured human exploitation. Relatively recent but rapid advances in the transportation of people and food and in fishing and hunting gear have in many instances deconstructed former relationships between humans, wildlife, and their environment. As well, some of the cultural, symbolic, and spiritual significance attributed to animals in earlier societies has been lost (Garibaldi and Turner 2004).

Marine bird populations in Atlantic Canada provide the medium with which we explore the socio-ecological restructuring of hunting and wildlife exploitation. As a window on these transformative patterns, we trace the changing dynamics and significance of seabird hunting in Newfoundland and Labrador, where this activity has been an integral aspect of coastal community subsistence for hundreds of years. Hunting has influenced marine bird populations and distributions, and community dependence on hunting has, in turn, challenged and shaped related government regulation and legislation. The Newfoundland murre hunt for thick-billed murres (or turrs *Uria lomvia*) and common murres (*U. aalge*) is a striking case in point. This hunt is the only legal non-aboriginal hunt of migratory seabirds in North America; its legality is essentially a term of Newfoundland's Confederation with Canada (Montevecchi and Tuck 1987).

Patterns of marine bird hunting changed radically through the twentieth century as consequences of (1) increased food security and reduced need, (2) patterns of exploitation, (3) improved technology, and (4) conservation science and policy. Our analysis of marine bird exploitation focuses on murres, common eiders (*Somateria mollissima*), and other seabird species, including the extinct great auk (*Alca impennis*). We explore how key influences evolved from antecedent conditions and how they affected avian populations and restructured seabird hunting in Newfoundland and Labrador. We assess how exploitation has influenced human food security, as well as efforts to protect and enhance marine bird populations.

EARLY HUMAN INTERACTIONS WITH SEABIRDS

The ornamentation, jewellery and carvings of aboriginal peoples reflect the significance of the animals that they exploit (Figure 7.1; see also Garibaldi and Turner 2004). Seabirds, especially extinct flightless Great Auks, held considerable significance for prehistoric Maritime Archaic People and more recent but also extinct Beothuks (McGhee and Tuck 1975; Tuck 1976; Montevecchi and Tuck 1987). In eastern North America's largest prehistoric cemetery (~3,000 years before present [B.P.]) at Port aux Choix on Newfoundland's Northern Peninsula, Maritime Archaic People interred great auk beaks in human burials, apparent indications of the esteem with which they regarded these birds (Tuck 1976).

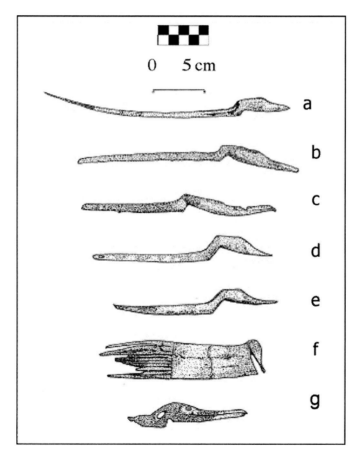

Figure 7.1. Avian effigies on pins and combs uncovered at the Maritime Archaic cemetery in Port aux Choix on Newfoundland's Northern Peninsula (Tuck 1976). Some are indicative of particular species and groups: loons (a, d, e), and mergansers or cormorants (crooked necks, hooked beaks—b, c, f, g). Figure drawn by M. Tuck (from Montevecchi and Tuck 1987)

Beothuks enjoyed great auk eggs, desirable for their large size (three times heavier than murre eggs), large yolks, and durability allowing for transport (Montevecchi and Kirk 1996). To get these eggs, they paddled ocean-going birch-bark canoes (Figure 7.2) across fifty kilometres of treacherous north Atlantic waters to the massive auk colony on the small rocky outcrop of Funk Island off the northeast Newfoundland coast (Cartwright 1792; Howley 1915; Lysaght 1971). Owing to the risks and dangers involved in reaching Funk Island by canoe and to the relatively limited egg harvest that they could make, the Beothuks' impacts on great auks likely posed no long-term population effects.

Figure 7.2. A Beothuk birch bark canoe designed with high mid-sides, bow and stern that may have made it more stable in rough seas. Reproduced by permission of the Department of Tourism, Culture and Recreation - Government of Newfoundland and Labrador. Artwork by David Preston Smith.

EUROPEAN PERSPECTIVES OF NORTH AMERICAN WILDLIFE AND ITS CONSEQUENCES

Wildlife abundance was extraordinary during early Europeans' ventures into the New World. In the Norse saga of Karlsefni dating back to 1007, eiders are noted to nest so densely on islands near Vinland that it was difficult to walk about without breaking eggs (Peters and Burleigh 1951). In 1497, the explorer Giovanni Caboto (John Cabot) encountered an extraordinary abundance of cod and other fish and wildlife, opening the way for the eventual European settlement of Newfoundland and Labrador. Correspondingly, perspectives of wildlife abundance emphasized a false perception of their limitlessness in the face of human exploitation.

Early Europeans who visited North America hunted seabirds, and the large, flightless great auks provided them with an accessible source of fresh, nutritious meat rich in protein and fat (Montevecchi and Tuck 1987). These birds, and especially those on Funk Island were vital to the food security of early European mariners who, following arduous north Atlantic crossings of a month or longer, were often nutritionally stressed and protein deficient. European fishers and settlers continued to exploit the auks for food and bait.

The great auks were so abundant that mariners also exploited them as navigational markers of the New World fishing banks (Montevecchi and Tuck 1987). In what appears to be the first population estimate of seabirds in North America, Sir Richard Whitbourne (1622) wrote about the great auks' "infinite abundance . . . divinely provided for the benefit and sustenance of man" (Figure 7.3). Unbeknownst to Whitbourne, he had also penned the species' eventual epitaph.

Figure 7.3. Great auks (*Penguinis impennis*), the last flightless birds of the Northern Hemisphere, extinct from over-harvesting

Overkilling appeared to be a *modus operandi* for survival. In his Labrador journal, Cartwright (1792) noted that he had killed eight polar bears in a single day and 1,500 eiders during migration from mid-April to May 1770. By the late nineteenth and early twentieth centuries, such excesses had nearly eradicated common eiders from the northeastern seaboard of North America (Goudie et al. 2000).

In North America, hunting practices and subsequent legislation surrounding them developed in New World ways. Actions were essentially socialistic antitheses of European laws that entitled land-owners strict control over access to wildlife (Lund 1980). Public domain and unchecked free access to wildlife were the unwritten laws of the land and sea. Yet these practices and perspectives did not apply to the aboriginal inhabitants, whose land rights were ignored or seized.

Public ideals ran counter to efforts at conservation. Such restrictive measures were perceived as impositions that would, as in the Old World, place access to wildlife under elitist rather than egalitarian control. The new settlers took an essentially adversarial approach in "conquering" the wilderness in which they found themselves. Within just three centuries of unrestrained exploitation, the short-sighted nature of the new North American culture of exploitation was all too evident (Cartwright 1792; Mowat 1982; Montevecchi and Tuck 1987).

The over-killing of great auks provides a telling case such that these birds have come to be regarded as a global icon of the need for conservation. Great auks absorbed the uncontrolled "harvests" directed at them for about two and half centuries. They could not, however, rebound from over-exploitation in the species' largest colony on Funk Island during the late 1700s. Commercial crews from nearby coastal communities camped on Funk Island, corralled the flightless auks, and slaughtered them excessively. Rather than for food, the auks were killed for their down and pin feathers that were used for bedding in mattresses and quilts.

Even at the time, Cartwright (1792) warned of the dangers of this unrestrained killing. Great auks, like many highly social marine (and terrestrial) birds and other wildlife, required large, robust populations to ensure their viability. The kills during the late 1700s fragmented and pushed the population below minimum viable levels and beyond resiliency (see Courchamp et al. 1999). Great auks were extinct by the early 1800s (Montevecchi and Kirk 1996). Because human over-exploitation caused this extinction, it is not completely facetious to imagine what populations of flightless birds might have done for current day ecotourism ventures on the eastern Canadian coast, as has been the case with penguins in the Antarctic.

Other hunted birds, including the Labrador duck (*Camptorhychus labradorius*) and Eskimo curlew (*Numenius borealis*), also fell within the aim of market and commercial hunters. They soon followed the great auk's fate (e.g. Gollop et al. 1986). Colonially nesting seabirds such as murres, puffins, terns, and eiders that concentrate in breeding colonies were particularly vulnerable to over-exploitation.

SEASONAL CYCLE OF HUNTING AND HARVESTING

To secure year-round food supplies, residents of coastal human communities, particularly those in boreal and Arctic environments, relied on seasonal cycles of animal and plant availability (Freuchen and Salomonsen 1958). Seasonal hunts for seabirds and seals, and to a lesser extent ptarmigan (*Lagopus lagopus*, *L. mutus*) and hares (rabbits), provided coastal residents with needed sources of fresh meat. Besides seabirds and their eggs, abundant summer and early autumn food supplies included fresh fish, root vegetables (suited to short growing seasons and acidic soils) from family gardens, and a diversity of wild berries (e.g. blueberries, *Vaccinium* spp., partridgeberries, *V. vitis-idaea*; bakeapples, *Rubus chaemaemorus*; Omohundro 1994; Karst 2005). The primary seabird species hunted during winter included murres (especially thick-billed murres), dovekies (or bullbirds, *Alle alle*), northern common eiders (or shoreyers, *Somateria mollissima borealis*), long-tailed ducks (or hounds, *Clangula hyemalis*), scoters (or divers, *Melanitta* spp.), and razorbills (or tinkers, *Alca torda*) (see Montevecchi and Wells 2006 for local Newfoundland names of birds). In summer, shearwaters (especially greater shearwaters *Puffinus gravis*, and sooty shearwaters, *P. griseus*; collectively referred to as bawks on the north coast of Newfoundland and as hagdowns on the south coast and in Labrador), Atlantic puffins (*Fratercula arctica*), black-legged kittiwakes (or tickle-ace, *Rissa tridactyla*), common murres (or Baccalieu birds, *Uria aalge*), southern common eiders (or shoreyers, *Somateria mollissima mollissima*), and other auks and gulls were hunted. In spring and early summer, eggs of eiders, auks, and gulls were collected at nesting colonies (Figure 7.4; Montevecchi and Tuck 1987), and some families raised gull chicks with a few domestic fowl that were slaughtered in autumn.

In winter, most of the seabirds present migrate into the region from nesting areas in the Canadian Arctic and Greenland. In summer, birds at breeding colonies on coastal islands were exploited, as were shearwaters that arrived in early summer from high-latitude colonies in the South Atlantic (Brown 1986). Thus local hunts in Newfoundland and Labrador have had, and continue to have, ocean-scale implications for seabird populations. Seabirds were, and are, shot from small open boats, whereas eiders and other seaducks were hunted from headlands that were accessed by walking or by dog sled. Dogs and floating "dog jiggers" with hooks were often used to retrieve fallen ducks on the water.

Until and even after the introduction and spread of household electricity and refrigeration in coastal areas during the 1950s and 1960s, murres and eiders were the most available and nutritious sources of protein during winter. Some families that depended on eiders for food would consume between 300 and 400 birds each year (Chaffey 2003). Supplies were often depleted during winter, especially when sea ice moved birds out of an area. In such circumstances, spring hunts for birds (and seals) and egging were often life-saving exercises.

In some communities (e.g. on the southern Labrador coast) eiders were traditional table birds, whereas on the north coast of Newfoundland hunters took large numbers of both murres and eiders (see recipe below). Dovekies, small auks from Greenland, were taken in large numbers and often cooked up as "bullbird soup."

Roast Turr or duck recipe

2 or 3 turrs/ducks	2 cups bread crumbs
1 tsp. savoury	1 tsp. poultry seasoning
1/4 cup melted butter	1 chopped onion
1/2 tsp. salt	1/4 tsp. pepper

Clean birds and drain. Stuff birds with dressing and skewer. Place in a pan and prick bird so that fat will drain off. After 1 hour in the oven at 300 degrees, drain off fat from pan and add some water and chopped onion. Cook some hours until tender, covered. Remove lid for last half hour, to brown skin. Serve with partridgeberry jam or jelly (from Various authors 1997 [*Five Hundred Years of Newfoundland Cookery*]).

Much of the fresh meat and other food was preserved as stores for the winter, when resources were less abundant and less accessible. People relied heavily on preserved foods, such as, salted fish and salted and bottled meat. Murres, eiders, ptarmigan (or partridge), hares, and seals were preserved by cooking and bottling or by freezing in outdoor sheds for later consumption beyond periods when fresh game was available. Eggs were often preserved in buckets of briny water or in barrels of sawdust. Fresh and preserved murres and eiders were traditionally bartered for other foods and services in communities or sold (Elliot 1991).

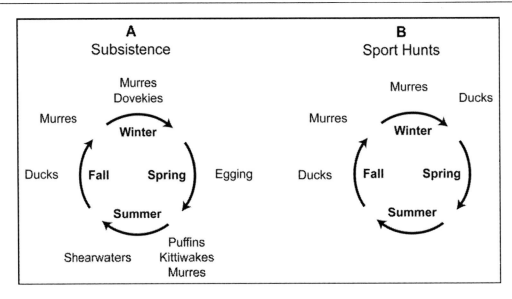

Figure 7.4. Past (A) and present (B) seasonal cycles of marine bird harvests

Fresh meat from seabirds was much more important for food security than as a supplement or a luxury in relatively isolated fishing communities. This quote from a Labrador hunter about eiders emphasized this necessity "There was times when . . . anything (food) else was pretty scarce . . . So . . . you used what you could get" (Chaffey 2003, 98). This need became very evident in the 1940s when the British Commission of Government in Newfoundland outlawed the hunting of seabirds during summer. This was the first prohibition of the hunting of shearwaters (bawks), favoured food birds that were abundant during summer fishing seasons (Montevecchi and Tuck 1987). Reaction from the coastal communities was vociferous and widespread. A poem, *The Shooting of the Bawks*, by Art Scammel from Change Islands on the northeast coast set the tone and became a rallying cry that engaged outport residents (see excerpts, below). Their reactions, emphasizing the need for the seabird hunt, resulted in a reversal of government policy. Such strong sentiments have carried over to the present and have influenced current hunting legislation in Newfoundland and Labrador.

> Excerpts from The Shooting of the Bawks
>   by Art Scammel
>   . . .
> No doubt our wise Commissioners will formulate a plan,
> To furnish fresh for everyone who lives in Newfoundland.
>   . . .
> For Mary dear I'll kill a bird in August, June or May,
> And if they put me in the pen, why there'll I'll have to stay,
> For men with children underfed, would rather far be sued,
> Than keep this bloody law that stops a man from getting food.
>   . . .
> Now bawks have got a fishy taste as everybody knows,
> But they make a better diet boys than either hawks or crows.
>   . . .
> There's many men in summer time who cannot buy salt meat,
> They have to trust for seabirds for something fresh to eat,

But if they keep this law that's passed, they will not get a taste
Of bawk or noddy, tinker, tur, and not a tickleace
(entire poem can be found in Montevecchi and Tuck 1987).

IMPROVEMENTS IN FOOD SECURITY: SHIFTING FROM SUBSISTENCE TO SPORT HUNTING

Following Confederation with Canada in 1949, the government of Newfoundland and Labrador actively pursued consolidation of coastal communities through a resettlement program, the establishment of roads into isolated regions and a cash economy. This social and economic restructuring led to a gradual shift away from subsistence living in these communities. Services and amenities, such as household refrigeration, helped to secure protein during winter.

Consequently but gradually over the next three decades, the motivation to hunt shifted from being one of need to one of recreation and sport (Lund 1980). It took much longer in rural communities than in the urban centres of North America for hunting to assume the role of the mainly recreational activity that it is today. Though they are no longer essential foods, murres and eiders are still savoured by many Newfoundlanders and Labradorians as refreshing alternatives to processed and frozen foods that dominate most North American diets. Many coastal residents still supplement their diets with marine birds because they enjoy them and because they are traditional meals (i.e. cultural comfort foods; see Garibaldi and Turner 2004).

Hunting seabirds from open boats during winter is still popular in coastal Newfoundland and Labrador. Hunting, however, is a waning activity in Newfoundland and Labrador and throughout Canada where licences for seaducks and seabirds are decreasing at more than 5% per annum (A. J. Gaston pers. comm.). This decline is expected to increase as rural coastal communities are depopulating, following the closure of the fisheries for northern cod and Atlantic salmon off eastern Newfoundland and Labrador in the early 1990s. Egging is no longer practised in insular Newfoundland, though some members of native communities in Labrador still harvest eider eggs (Chardine 2001).

The shift to sport hunting also set the stage for a proliferation of conservation legislation and regulations. Many of these changes sharply curtailed utilitarian practices, such as the illegal selling of birds in an underground economy and market hunting. "Game" species were given considerable attention and protection. The consequences for wildlife exploitation and protection were pervasive, and sport hunters, such as those involved with the Partridge Forever Society in Newfoundland and Labrador and with Ducks Unlimited, actually played and are playing major roles in helping to protect wild bird populations.

INFLUENCES OF TECHNOLOGICAL CHANGE

Technological development had, and continues to have, pervasive influences on the abundance, distributions, and even extinctions of many marine animals (Agular 1986; Steele et al. 1992; Hutchings and Myers 1994; Montevecchi and Kirk 1996; Kurlansky 1997; Burke et al. 2002; Myers and Worm 2003; Springer et al. 2003). A recent example of improved hunting technology and the decimation of seabird breeding populations in Greenland can be found at www.birdlife.org/news/2006/01/greenland.html. Complexity in marine food webs has been diminished (Pauly et al. 1998), leaving many animals vulnerable to further perturbation, such as those induced by climate change, and subject to severe population declines (Myers at al. 1995; Jackson et al. 2001; Stenhouse et al. 2002; Gasciogne and Lipicius 2004).

When eider hunters in Labrador began using snowmobiles to access coastal hunting sites (~1970), they could travel at least twenty kilometres further on hunting trips than when they travelled by foot. Many hunters considered that the greatest change in hunting capabilities occurred with the introduction of outboard motors in the late 1960s and 1970s (Table 7.1). Before extensive outboard motorboat ("speedboat") use in the 1970s, hunters used rowboats (punts) and trap skiffs during eider and murre hunts. Speedboats enabled hunters to move faster (up to forty-five to seventy kilometres per hour), cover more area, "chase" birds and "round them up" in coves, and to generally hunt more efficiently, killing more birds in less time. The use of speedboats drove eiders and murres away from community shorelines. The net result was that not only could hunters travel farther to hunt, their improved transportation efficiency created circumstances that required them to go further to find birds. An analogous situation in terrestrial hunting has occurred during fall ptarmigan hunts, in which all-terrain vehicles (ATVs) have increased the efficiency and coverage of hunters, leaving the birds little spatial refuge in many areas of prime habitat. About a decade or so after the introduction of outboard engines, new fibreglass boats allowed hunters to move through the winter sea-ice, greatly increasing access to birds at sea (Table 7.1).

In recent years, many hunters have begun towing their boats on trailers from bays on the northeast coast to the southeast coast of Newfoundland, as open seasons and bird movements shift around the island during winter. These changes in hunter capability and activity influence marine bird behaviour and distributions on the Labrador and Newfoundland coasts, generally causing them to move further from human settlements and further from near-shore feeding sites, as well as to become more wary and flighty (Chaffey 2003).

Table 7.1. Changes in marine bird hunting practices and technology during the twentieth century based on interviews with hunters

Subject	Early 1900s	1940s-60s	1970s	1980s	1990-present
No. hunters per capita	high	high	high-moderate	moderate	moderate
Boats	rowboats	trap skiffs/ inboard motors	wooden speedboats/ outboard motors	fiberglass speedboats/ outboard motors	fiberglass speedboats/ larger outboard motors
Shotguns	Muzzle loader	muzzle loader; break action	break action	bolt; pump action	semi-automatic
# eiders per family annually	?	50-400	50-300	50-300	10-60
Monetary cost of hunting	low	low	moderate	moderate	high

Sources: Elliot 1991; Chaffey 2003

Improved weaponry has also enhanced hunter capability. Before the early 1980s, many hunters used single-shot shotguns, such as a break action. These were rapidly replaced by pump action then semi-automatic shotguns that hold more cartridges and can be discharged more quickly. Federal hunting regulations stipulate that hunters using automatic rifles can only load three shells at once.

Hunting is only one source of human-induced mortality that seabirds and seaducks have to cope with in Newfoundland and Labrador. In the next section, we consider the others and the cumulative effects of these diverse sources of anthropogenic mortality.

CUMULATIVE MORTALITY

Hunting migratory marine birds helped sustain coastal communities, but hunting and egging in nesting colonies often carried severe consequences. The coastal breeding sites of many seabirds in Newfoundland and Labrador have been decimated since the arrival of Europeans (Montevecchi and Tuck 1987; Goudie et al. 2000). Fuelled by market demands for birds, eggs, and feathers for the millinery trade, the destruction of seabird colonies was widespread along the eastern North American coast (Forbush 1912; Gollop et al. 1986; Greenberg and Reaser 1995). Owing to public outcries in the U.S.A., the *Migratory Bird Treaty Act* (1916) was created to restrict the numbers of birds being killed and to protect vulnerable species. The legislation recognized the importance of international cooperation in protecting migratory species that moved across national borders.

Winter hunts also carried population consequences but these were less evident as migratory birds moved out of the area to distant colonies during the breeding season. The murre hunt still kills hundreds of thousands of thick-billed murres from the Canadian Arctic and Greenland annually (Elliot 1991). When cumulated with other sources of anthropogenic mortality from oil pollution and by-catch in fishing gear (Piatt et al. 1984; Montevecchi and Tuck 1987; Wiese and Ryan 1999; see also Wiese et al. 2004), these kills were assessed to be unsustainable (Elliot 1991). For example, estimates indicate that hundreds of thousands of seabirds including ducks are killed each year by illegal oily discharges from ships in the Northwest Atlantic (Wiese and Ryan 1999). Murres and other diving seabirds are also subjected to considerable mortality from entanglement and drowning in fishing gear, such as gill-nets (Piatt et al 1984; Piatt and Nettleship 1987; Montevecchi 2001). This mortality has become evident following the eastern Canadian ground-fishery closure in 1992 that resulted in the removal of many thousands of kilometres of gill-nets from the waters of Labrador and eastern Newfoundland (W. A. Montevecchi, unpublished data). Consequently, local breeding populations of diving seabirds are rebounding (Robertson et al. 2004) from this multi-decadal source of mortality. These increases do not however apply to the migrant murres and eiders from the Arctic that are targeted by hunters in Labrador and Newfoundland during autumn and winter.

Eider populations are being threatened and stressed by habitat losses associated with extensive mussel culture sites throughout the region. Mussel farms take over eider habitat sites and also at times attract the ducks to artificial food concentrations. The eiders and other seaducks are often viewed as competitors by mussel farmers who scare, chase, and at times kill seaducks at mariculture sites (Montevecchi 2001). As aquaculture ventures replace wild fisheries (Fischer et al. 1997), these anthropogenic pressures will increase.

THE EVOLUTION OF HUNTING REGULATIONS

The first legislation for the protection of birds in what is now Canada was enacted by Newfoundland's colonial government in 1853 (Forster 1978). The *Act for the Protection of the Breeding of Wild Fowl in this Colony* stipulated total protection including the taking of eggs from 10 May to 1 September, with a maximum fine of £20, a huge sum at the time. This Act was replaced in 1859 with a more comprehensive one that made it specifically illegal to collect eggs from Funk Island and protected partridge (ptarmigan) and snipe (*Gallinago gallinago*) from 10 May to 10 August. Food security was also considered, as these acts exempted hunting out of necessity: "Nothing in this Act shall extend . . . to any poor Settler, who shall kill . . . Wild Fowl, for his own immediate consumption, or that of his family." The traditional right of coastal residents to procure birds for food (even during closed seasons) was upheld in one way or another as long as Newfoundland was a British colony and even after Newfoundland joined Canada (Montevecchi and Tuck 1987). The 1859 Act was amended in 1863 to prohibit the use of guns on Sundays, a prohibition that remained in effect through 2005, when it was modified to accommodate recreational outfitters and some "big game" hunters.

Following Newfoundland's confederation with Canada in 1949, federal hunting regulations and the Migratory Birds Convention Act, an international treaty with the U.S.A., soon came into play in Newfoundland and Labrador. The treaty between Canada and the United States for the protection of birds prohibited the hunting of migratory seabirds. Yet, because the winter murre hunt was related to family food security and also carried cultural significance (see Garibaldi and Turner 2004), the treaty was amended to allow any Newfoundland and Labrador resident to shoot murres for food—essentially making the hunt a term of confederation. It was illegal to sell murres, but without enforcement it was difficult to prevent it. Selling resulted in more murres being taken than were needed to secure family provisions. Also due to open hunting season, increasingly more adult birds were being taken later in the spring when they returned to breeding sites. Like many seabird species, murres are long-lived with delayed maturity and lay only one egg per year. Hence, their populations are highly sensitive to slight changes in adult mortality.

The cumulative mortality associated with the murre hunt, oil pollution, and entrapment in fishing gear appeared sufficient to make the hunt unsustainable (Elliot 1991). The first federal regulations to control the hunt were imposed in 1993. Daily bag limits were set at twenty birds per hunter, with possession limits of forty birds. Hunting seasons opened and ended earlier in Labrador, opening and closing progressively later in hunting zones along the north, east, and south coasts of Newfoundland. Murre hunters were by and large in agreement with the new regulations as many felt that the massive annual harvest of murres and the common practice of selling birds was making the hunt unsustainable (Elliot et al. 1991). The application of hunter or local ecological knowledge (LEK) has proved to be a valuable tool in wildlife management, often as an indicator of a need for further scientific scrutiny in situations where over-exploitation and disturbance have gone undetected by scientists and regulators (Gilchrist et al. 2005).

In the 2003–04 season, all hunters were required to purchase and be in possession of a Migratory Game Bird Hunting Permit and Habitat Conservation stamp while hunting murres. Revenues raised by these "wildlife taxes" are to be used to support research and conservation. Mandatory hunting permits will also improve the ability of regulators to track the numbers of birds taken annually during the hunt. Hunter surveys indicate that substantially fewer murres are being killed as a result of these hunting restrictions (Figure 7.5).

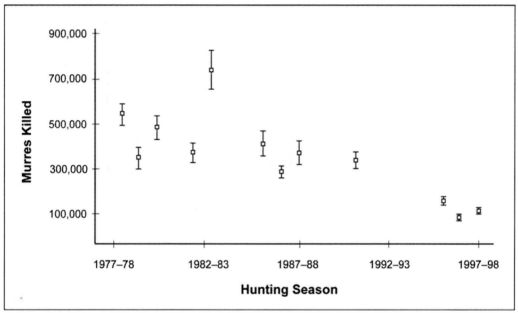

Source: Chardine *et al.* 1999; see Elliot *et al.* 1991 for methods

Figure 7.5. Mean (± standard error) annual murre harvest estimates from hunter surveys beginning in 1977

NON-CONSUMPTIVE USES OF BIRDS IN ECOTOURISM

With food security no longer an issue in most coastal communities, with greater restrictions on hunting, and with waning hunting activity, many people have turned to non-consumptive uses of marine birds and mammals. These activities include bird- and whale-watching, wildlife photography. Many participants focus on the ecological, ethical and aesthetic perspectives and values of wildlife for its own sake. Ecotourism is a rapidly growing economic sector in Newfoundland and Labrador. Unlike other highly localized economic drivers (e.g. a mining site), ecotourism revenues spread widely into rural and coastal communities, helping to support sustainable developments.

Ecotourism activities, like consumptive ones, also carry costs for wildlife, in terms of disturbance and displacement (Wang and Miko 1997). Comprehensive efforts are needed to effectively prevent these (Isaacs 2000). Emphasis on key "game" species is being replaced by conservation perspectives that often focus on charismatic megafauna (e.g. puffins and whales) and rare species (e.g. harlequin ducks *Histrionicus histrionicus*; ivory gulls *Pagophila eburnea*) and also embrace biodiversity and the protection and inter-connectedness of habitats.

CONSERVATION RESEARCH, EDUCATION AND HABITAT ENHANCEMENT

Our *Coasts Under Stress* research focused on eider hunting and applied local ecological knowledge (LEK) to help understand historical and current trends in hunting and eider populations in southern Labrador. We worked with hunters to decipher long-term patterns and to develop conservation perspectives for robust eider populations. This research developed into a collaborative project with hunters and was co-sponsored by Environment Canada and Ducks Unlimited. Nest shelters were positioned and built on eider-nesting islands in St. Peter's Bay (Figure 7.6). We developed and distributed information about the project with feedback questionnaires to all households in the local communities before project initiation. Throughout the course of our research, educational programs in the schools and for the general public complemented these initiatives to help integrate our activities into community traditions in order to enhance, promote and modify ongoing conservation activities and policies.

Figure 7.6. Eider nest shelter in St. Peter's Bay Photo: J. Coffey.

Synopsis

Hunting seabirds to secure food was a vital aspect of life in coastal communities in Newfoundland and Labrador through about the mid-twentieth century. These early hunts were sustainable because coastal settlements were small and widely dispersed, and because the technology at the time limited the spatial scale and killing capacity of the hunters. Subsequent improvements in engines, boats, and weaponry during the last half of the nineteenth century greatly increased the efficiency and foraging range of the hunters often resulting in overexploitation and disturbance. As the introduction of household electricity and the importation of food secured family provisions, hunting incentives shifted from necessity to recreation. Seabird hunting has been waning with the depopulation of coastal communities. Yet owing to cumulative human influences throughout marine ecosystems (e.g. Jackson et al. 2001; Burke et al. 2005; Montevecchi 2006), the conservation of marine birds and their habitat is compelling environmental concern. The non-consumptive exploitation of wildlife through ecotourism and the integration of local environmental issues and values in ongoing educational and public programs can be of great benefit in helping to sustain the viability and integrity of coastal communities.

Acknowledgements

Our *Coasts Under Stress* research was supported by a Major Collaborative Research Initiative (MCRI) grant sponsored by the Social Sciences and Humanities and Natural Sciences and Engineering Research Councils of Canada (R. Ommer, Principal Investigator). We are grateful to the hunters, fishers, teachers, and residents who made input and collaborated with us. We thank Juliana Coffey and Jodi Baker for research assistance in Labrador, where many people helped us, most especially Jack and Joy Rumboldt. We thank Barbara Neis for help with the research, and Editors Chris Parrish, Nancy Turner, and Shirley Solberg for constructive comments in the preparation of this chapter.

In: Resetting the Kitchen Table
Editors: C. C. Parrish et al., pp. 115-127

ISBN 1-60021-236-0
© 2008 Nova Science Publishers, Inc.

Chapter 8

FOOD SECURITY AND THE INFORMAL ECONOMY

Rosemary E. Ommer[1], Nancy J. Turner[2], Martha MacDonald[3] and Peter R. Sinclair[4]

[1]*Coasts Under Stress* research project; SSHRC Grant Facilitator, University of Victoria;
History, University of Victoria and Memorial University of Newfoundland, Canada
[2]Environmental Studies, University of Victoria, Canada
[3]Economics, St. Mary's University, Canada
[4]Sociology, Memorial University of Newfoundland, Canada

ABSTRACT

First Nations coastal communities, which have existed on the coast for millennia, have a holistic vision of humans and the natural world, in which people and place are eternally interwoven. For other traditional communities, especially those on the East Coast where European settlement took place more than two centuries ago, many people had not so much a seamless embeddedness of community and place, as a learned and cherished relationship with the places in which they live. For both, economy was based on the extraction of resources for sustenance, but on the East Coast that was married to extraction for a wider marketplace accessed through merchants, a situation that evolved into a formal industrial capitalist economy supported by law and the state, and an informal sector, which is a group of economic activities that operate outside the formal legalized structures of a nation's capitalist marketplace. This chapter first considers the history of First Nations on the West Coast, and settler communities on the East, to situate present day informal practices within the historical experience of food provision, culture, and lifestyle. It then turns to the present day, to examine the economic role and cultural meaning of the informal sector on both coasts, in order to identify the ways in which such informal practices secure sustenance and how altered they have been by restructuring. The basic purpose is to understand whether or not such practices can survive, and contribute to the life of coastal communities and, indeed, the wider society of which such communities are an important part.

INTRODUCTION

This is a book about food security and its relationship to the social-ecological systems in which coastal communities exist. Indigenous coastal communities which have existed for millennia, have as part of their culture a holistic vision of humans and the natural world, in which people and place are eternally interwoven. For other (settler) communities, especially those on the East Coast where white settlement took place more than two centuries ago, many people have not so much a seamless embeddedness of community and place, as a learned and cherished relationship with the places in which they live. For indigenous peoples, a relationship of reciprocity with both nature and other First Nations has been the ideal type of relationship for their economy and culture. For them, as for other tribal communities in Europe and elsewhere, economy was not seen in capitalist wage-based terms, but rather in the way that the ancient Greeks saw it when they talked of "oecumene" (household)—that part of cultural life that sustained the household and the clan or lineage, and which (as extended family) was a broader concept of the household—an extension of it. In non-First Nations coastal communities in Canada, there initially existed an economy based jointly on the extraction of resources for sustenance, and for a wider marketplace that people accessed through the merchants for whom they worked. In later years, that evolved into a formal industrial capitalist economy supported by law and the state, and an informal sector, which is a group of economic activities that operate outside the formal legalized structures of a nation's capitalist marketplace.

The informal sector is based in community or family reciprocities (for a detailed discussion, see Ommer and Turner 2004, 127–57) and involves a sophisticated utilization of the non-market products and services associated with occupational pluralism. These consist of

> . . . the utilization by a community or indigenous group of a range of geographical locations and ecological niches which, taken together, provided year-round sustenance, both in terms of foodstuffs and other necessities of life. They are, therefore, both place-specific in operation, and rural, constituting what we might call 'ecological pluralism,' and they remain a vital part of the rural resource-based coastal communities of eastern and western Canada (Ommer and Turner 2004, 128).

In this chapter, we consider such informal economies with respect to the way in which they contribute to food security, as defined at the beginning of this volume. In so doing, we draw on research findings from several parts of *Coasts Under Stress* (Ommer and team, in press, for an overview of the findings of the whole team), and also from a growing literature on food and First Nations *genres de vie*, and food in informal non-aboriginal systems. We look first at the history of First Nations (indigenous) communities on the West Coast, and settler communities on the East, to situate present day practices within the historical experience of food provision, culture, and lifestyle, as it has been analysed by ourselves and others. We then turn to the present day, to examine the economic role and cultural meaning of the informal sector in our study areas on both coasts, as these communities seek to reposition themselves in a rapidly changing world. Our fundamental purpose is to identify the ways in which such informal practices secure sustenance, to see how altered these practices have been by restructuring, and to ask whether they can survive and make a meaningful contribution to coastal communities and to the wider society of which such communities are an important part. It is the sustenance that was provided by these practices even in "hard times" that led Newfoundlanders in the 1930s Depression, for example, to boast that "you'll never starve here." However, we will show that under current conditions of major fish stock failures (East Coast) or declines (West Coast) and dwindling informal economies, the food security that was once part of rural life on both coasts is now endangered.

A BRIEF OVERVIEW OF THE HISTORY OF FOOD SECURITY ON BOTH COASTS

As Ommer and Turner (2004), among others, have shown, informal economic activities and strategies have always been important for community survival on the coasts. The underlying resilience of rural communities, both in the past and today, has always been drawn from their not-very-visible and now seriously threatened "informal" economic structures, which, we maintain, are not only economic in nature but also social and cultural, including ethical "rights" and obligations, with supporting institutional mechanisms such as the potlatch on the Northwest Coast. These arrangements are rooted in the fundamental interactions among rural community subsistence, local resources, demographics, ecology, and economy. In the past they relied totally on an exchange of equivalent goods and services; today, they sometimes involve cash as a medium of exchange. Reciprocation is, however, often "in-kind"—shared baby-sitting, one skill proffered in exchange for another ("I'll help you fix your house; you help me with the harvest")—and thus the informal sector works locally, with economic exchanges being primarily to enable household and community survival. The creation of surplus for wealth or profit is not of concern. Even now, under conditions of relative poverty in many coastal communities, informal economies are highly effective, not least because they are supremely pragmatic. They provide community safety nets, play a vital role in social cohesion and often function in stewardship of the environment in which they are embedded.

Informal economies are also expressions of living cultures, adaptable to change and capable of the great flexibility which has in the past rendered rural communities extremely resilient. In the face of social, economic, and environmental global restructuring, however, the traditional resource-based economy has been increasingly abandoned or downgraded and its informal counterparts circumscribed, thus threatening the quintessential community capacity to adapt. This is one reason why small communities are feeling increasingly marginalized and neglected, and it intensifies their impoverishment under post-industrial restructuring.

In Newfoundland in the eighteenth century, migrants were brought to the shores of the distant island colony to fish for merchants from the west country of England and the Channel Islands under indenture agreements that required them to remain on the coast over two summers and one winter. Over time, this migratory fishery gave way to one in which a settled labour force worked for merchants under a truck system in which the merchant exchanged fishing services for essential supply goods which could not be created on the coast. The arrangement worked to the advantage of both fisher and merchant in that it provided fishing families with access to supply goods and the marketplace for fish, while the merchant received the fish which he traded in an international marketplace, in exchange for higher value goods, the profits from which returned to his home base in southwest England or the Channel Islands (Ommer 1990a, 1990b, 1991).[1] In the process, settlers developed a system of occupational pluralism in which they provided for themselves those necessities they could produce (food, some clothing), thus purchasing as little as possible on credit against fish in the merchant store. The flexibility and adaptive capacity generated in this strategy resulted in the ecological pluralism of the Newfoundland "outport" (a term that derives from colonial days, when fishing stations outside the main port were called outports) and survives today in altered and abridged form as the informal part of the outport economy. It has remained a feature of households and extended kin-groups in rural areas, where it is found—now as in the past—operating as a combination of activities which exploit a range of ecological niches made accessible through the limited nature of the fishing season. Thus, small gardens produced local food (carrots, potatoes, cabbage, and sometimes a goat and some poultry for meat, milk, butter, and eggs), surrounding land provided wood for construction and fuel, and meat obtained from hunting, while fruits were picked from the hinterland "berry barrens" (Figure 8.1).

In addition to the resource-based seasonal round, the informal economy also incorporated, then as now, household exchanges of what might be termed reproductive or "caring labour" which sustained families and communities. However, with greater material prosperity following Confederation, state transfer payments and the development of a more diversified economy, subsistence work became less vital merely to survive although, even in relatively secure economic circumstances, subsistence work remained widespread (Omohundro 1994). The Newfoundland informal economy is just one expression of a mode of subsistence existence that applies to many communities across Canada, and elsewhere (McCann 1982; Harris and Warkentin 1974). The marginal nature of individual resources taken in isolation had likewise engendered the seasonal rounds found among indigenous peoples in the region: Beothuk and Mi'qma'q, for example (Ommer 2002b, 22–25).

Figure 8.1. Blueberry picking near Twillingate, Newfoundland. Photo: David Renfroe

---

[1] See Ommer, 1990b, especially Introduction and pp. 49-72, 86-167 for various aspects of the Newfoundland merchant system, including the transition from migratory to resident fishery in our field area.

On the West Coast, what we now speak of as the informal economy was, until colonial contact, simply the indigenous economy, which operated through the sanction of the leaders of all First Nations, both within and across communities. However, the imposition of European market-driven and colonial economic systems upon these traditional indigenous systems created both social and ecological imbalances in the existing indigenous systems. For the First Peoples, participating in the activities of the new economic regime was seen as a matter of survival. Nuu-chah-nulth hereditary chief Earl Maquinna George recalled how, when he was a young man, he left Ahousaht to try to make a living at the big sawmill in Port Alberni, Vancouver Island: "I was put on the night-shift, piling the heavy timbers as they came out from the sawmill. I learned how to operate and manoeuvre them by using a levering tool called a peevee. As a youngster, the work was heavy, the hours were long, and the noise was terrible. Throughout the night I could hear the saws screaming in my ears. But I had to make a living" (pers. comm. to NT, 1999). Others found ways of extending and modifying their traditional activities into the wage or market economy, through seasonal work in canneries and fruit harvesting, or through home-based production of art and clothing.

In the case of many women, basket making was an activity that converted readily from domestic application of products to forms of exchange, within the informal more often that formal economy (Turner 2003a). Such activities usually allowed women opportunities to care for their children, and in fact, children often participated in production even at a young age. Many contemporary elders remember helping their parents and grandparents in food production activities. For example, while their parents, Mabel and Herbert Ridley, were away working in the canneries during the summer, Belle Eaton and Colleen Robinson, now Gitga'at elders living in Hartley Bay on the north coast of British Columbia, along with their siblings, used to stay with their grandmother Lucille Clifton and help her pick berries and wild crabapples, smoke salmon, and prepare salmon egg "caviar" and other food for winter. Their own nutrition was enhanced by these activities, as they would snack on the berries as they picked, as well as partaking of the food they had helped to preserve later in the year. This kind of practice continues—see Figure 8.2.

Figure 8.2. Ethan helping to pick berries, Hartley Bay, British Columbia. Photo: Nancy Turner

As a nine-year old child in the early 1900s, the late Margaret Siwallace of the Nuxalk Nation, Bella Coola, used to go to the nearby cannery where her aunt worked as paid labour. While her aunt cut the salmon and packed it into cans, Margaret would collect the discarded fish heads and tails. These, she took home and, gathering driftwood from the estuary for fuel, she smoked these otherwise "waste" products, and thus provided her family's food supply for the winter (pers. comm. to N.T. 1984).

In terms of the concerns of this chapter, then, the historical experience of food provision for coastal communities on both coasts was tied into their various cultures and lifestyles and provided the wherewithal for household existence alongside (after European contact) the formal economic structures of the day, in both mercantile and industrial times.

THE INFORMAL SECTOR TODAY

The informal sector survives on both coasts, and continues to be vital to the well-being of households and communities. Our research in coastal BC did not highlight the informal sector except with regard to First Nations, although we suspect it is alive and well in virtually any small outlying community. For west coast indigenous people, having access to the productive places and resources within their traditional territories was and is paramount for them to be able to maintain their health and resilience through the informal economy. This access has been seriously eroded in recent years and, moreover, there has been serious loss of biodiversity and, with that, loss of some culturally and nutritionally important species. This is a direct result of the imposition of colonization and the capitalist economic system. Out of a long list we offer, as examples only, loss of cranberry bogs which were drained for agriculture or highway construction, of creeks where salmon and trout formerly abounded, of tidal flats where people used to cultivate their root vegetables, of cedar trees on whose bark and wood people relied, and of yew trees (*Taxus brevifolia*) which were needed for their wood and for medicine. Moreover, even where resources still exist, First Nations have lost access to many of them as the forests and fisheries of the West Coast have been leased, sold, or given as quotas, to corporate interests, thereby further exacerbating the problem of cultural and nutritional loss which is at the base of most land claims today. Yet, despite this, there remain elements of an ongoing informal economic system with both resource and cultural components, which continues to function both within and among aboriginal families and communities, and between them and their non-aboriginal neighbours.

Many coastal people still gather seaweed and fish for food, which they process themselves, saving some for exchange. Those who help prepare the food receive a share, which is particularly important for communities where there has been serious loss of access to resources, resulting in low incomes and employment. We have seen this functioning at seaweed and salmon camps, in food fishing, and processing the catch (Hood and Fox 2003; Turner 2003b). Partaking in the harvesting of traditional food is also part of an exchange network in which locally plentiful resources are exchanged for those which are scarce there. Many different social institutions are used to fulfill this function, of which potlatching is the most well-known: potlatch gifts distributed to invited guests include many nutritious food items—jars of smoked salmon, wild berry jams, jarred soapberries, dried seaweed, and oulachen grease—as well as basketry, sewn blankets and vests, and other works of art. The process is reciprocal and, in time, the hosts will themselves receive similar gifts. Sharing also occurs when people have helped process food, or when people meet at important local gatherings. Indeed, this kind of exchange of nutritious foodstuffs is standard practice among west coast First Nations and is one important way in which peoples' health and well-being are maintained, even under the stress that they experience as a result of severe restructuring. It is part of the cultural heritage of First Nations, and is one example of the complex system of knowledge and practice (traditional ecological knowledge) that speaks to a society in which environmental monitoring and stewardship was simply part of daily life. Such knowledge, although very place-specific and spoken of with a local language and vocabulary, is nonetheless representative of an approach to people and the environment in which they are embedded which respects biodiversity and the inter-relatedness of all living things. It is critically important to human survival in a planet where homogeneity leaves us vulnerable to environmental and cultural collapse (Berkes 1999; Turner 2005).

What has become of the informal economy in Newfoundland and Labrador in the recent context of social, economic, and environmental restructuring? Does it continue to maintain household livelihoods and food security and sustain local culture? Despite the importance of the cash economy in recent years, subsistence activities and support networks have remained critical components of rural livelihoods in general, and the two are interdependent—subsistence activities stretch scarce dollars, and cash is needed to fuel the subsistence pump. Some subsistence activities also have a cultural and recreational value and are not just engaged in for livelihood purposes (see Montevecchi et al., Chapter 7, this volume). Such was clearly the case on the Great Northern Peninsula just prior to the 1992 moratorium. Earlier research by Felt *et al.* (1995) demonstrated that over 70% of 250 surveyed households regularly supplied themselves with a wide range of subsistence products. Moreover, they did so in large part to meet cultural expectations. That is why the correlation between household income and participation in the informal sector was not significant.

In our research for the period since 1991, we found that many traditional subsistence activities were alive and well. Our interviews on household strategies in Newfoundland and Labrador produced considerable information on particular subsistence activities, who engaged in them, and the respondents' perception of changes in dependence in 2001 compared with 1991. Depending on the activity, we have information from at least forty, and up to fifty-three, households covering three areas: White Bay South, Hawke's Bay/Port aux Choix, and the Labrador Straits. In our discussion of these activities, unless there is a major difference across the three areas, we shall combine these data (Solberg et al. Chapter 11, this volume).

Table 8.1 shows twenty-three subsistence activities in order of household participation in these communities. More than half of the households reported members who cut wood, hunted moose and rabbit, picked berries, fished for cod, and/or grew potatoes or other vegetables (see Figure 8.3). Most households in the Labrador Straits hunted for birds (turrs and partridge) (see Montevecchi et al., Chapter 7, this volume) but few did so on the Northern Peninsula. This was the only sharp difference in participation among the areas. Most wood was cut for fuel rather than for building. Compared with earlier research in 1988 by Felt and Sinclair (1992), the gender-based division of labour in these activities showed signs of weakening, as both partners, or a combination of parents and children, engaged in the main subsistence activities. Of the five most common tasks, it was only in cutting and hauling wood that the male spouse tended to work alone (54.8%) rather than with more widespread household involvement. This situation is paralleled on the West Coast, where, for example, the Gitga'at harvesting of seaweed, formerly exclusively undertaken by women, is now commonly an activity of both men and women (Turner 2003a).

Table 8.1. Subsistence Goods, Great Northern Peninsula and Southern Labrador, 2002

Type of good	Number of households responding	Percentage engaged in activity
Wood	48	87.5
Moose	53	79.0
Berries	41	68.0
Codfish	53	66.0
Potatoes	54	55.5
Veg/Fruit	57	53.0
Rabbit	53	51.0
Trout	53	49.0
Turrs	53	34.0
Seals	53	30.0
Turnip	53	24.5
Salmon	53	23.0
Cabbage	53	23.0
Partridge	54	17.0
Greens	53	15.0
Carrots	53	15.0
Beets	53	11.0
Caribou	38	10.5
Seafood	53	9.0
Sea Ducks	53	9.0
Onions	53	9.0
Broccoli/Cauliflower	53	7.5
Strawberries	53	5.5
Rhubarb	53	4.0

Given the entrenched cultural value of some activities (food fishing, hunting, for example), we would not expect a close correlation between subsistence work and economic decline. That said, it is possible that the impact of restructuring after the cod moratorium could have encouraged greater involvement in the informal economic sector to replace lost opportunities in the formal economy; or, as observed on the West Coast, restructuring may have eroded opportunities in the informal, as well as the formal economy. A majority of households reported that their dependence on subsistence activities was about the same as in 1991, whereas 25% claimed more participation (see Figure 8.4). Overall, the statistical evidence supports our view that informal subsistence activities are widespread over the study area, but not impacted in a consistent way by restructuring; there are some forces acting to increase subsistence activities and some to decrease them.

Figure 8.3. Roadside garden in the Great Northern Peninsula, NL. Photo: Peter R. Sinclair

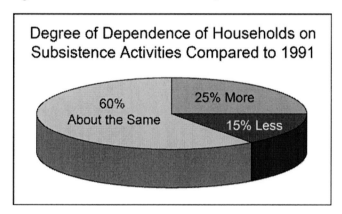

Figure 8.4. Comparative dependence on subsistence activities, Newfoundland, 1991–2004

We found that restructuring has undermined some traditional subsistence opportunities. Environmental changes threaten subsistence activity in some areas and resource management policies further limit people's traditional access to wood and fish, often to their consternation and discontent with perceived inequities. One local fisher commented, "There is companies here catches, you know with their own trawlers and that, goes and catches more fish in one day than all Newfoundland would take probably . . . well one boat probably like take more than all Newfoundland would take for to put their fish up for the winter." Restructuring has also affected the cash available for "do-it-yourself" projects. For example, in the 1990s the availability of cash from employment insurance benefits or The Atlantic Groundfish Strategy (a program designed to compensate and retrain fishers and plant workers who lost employment due to the groundfish moratoria) enabled people to buy some of the start-up materials necessary for activities such as building and repairing homes. Since that time, cash resources to support subsistence activities appear to have become more constrained. For example, one man explained that he burned oil now rather than cut wood, "because for one thing [to get wood] you got to have a truck; got to have a chainsaw. You got to have a skidoo. And you got to have gas. And then you got to go then a long way then, you got to go way up there . . . ”

Economic restructuring has also facilitated certain subsistence activities, however, sometimes turning them into cash generating opportunities. Some people turn to such activities when other more formal livelihood options decline, as well as just to give themselves something to do. In the Labrador Straits, for example the development of a local business producing wild berry jams and syrups has enabled many people to supplement their incomes through picking bakeapples (also known as cloudberries—*Rubus chamaemorus*; Karst 2005). While this activity has gone on for generations, its economic meaning has changed. An ex-fisher noted that bakeapple picking was something he couldn't do before, because it conflicted with fishing time. “But now I can't go fishing, so I try to do something else to get a few extra dollars.” Twenty-five percent of families said they use subsistence production to generate extra income throughout the year to help pay bills and make purchases they would otherwise be unable to afford. While not new, this is one place where people can attempt to address income shortfalls.

Census statistics obscure the complex interactions around subsistence activities, which often link various households. An example from southern Labrador is instructive. A middle-aged couple lived with two of their adult children and were active in the informal sector, partly for subsistence and partly to earn unofficial income. Asked about subsistence activities, John replied that they were mainly active in picking wild berries, partridgeberries (also called “red berries” or lingonberries—*Vaccinium vitis-idaea*) from their local area and strawberries from the Northern Peninsula. Sometimes they sold the strawberries. However, moose meat was given away to others, which is a common pattern of distribution and integration among households. Both spouses took part in hunting moose and rabbit, but they rarely ate any themselves. “Sometimes we go out (hunting)—we was out last year, me and her. We enjoys it, but we don't eat it.” Much of this food was distributed to senior citizens. People still value going on the land. Subsistence activities are not always done for immediate personal gain— though what goes around may come around.

Bill and Daphne, manual workers in southern Labrador, illustrate an important feature of subsistence work throughout the area. They built their own home, which contributes substantially to their material quality of life.

> We built it on our own and we owned it from the start. Oh yes, we done a lot (renovations) . . . we built it off and on. When the fishing was over, we took a couple thousand dollars and we did it by bit . . . The cupboards went in last fall and the flooring done. We are putting in a new bathroom down stairs—that's all. He does all that himself, the construction work around the house. He does it all except we had to hire someone for the cabinets.

It is generally believed that informal networks in the past were important as a means for households to draw on labour of others or obtain subsistence supplies as needed. Specific exchange was usually not expected (Felt and Sinclair 1992). For 1988, Felt *et al.* (1995) reported that almost 82% of households were linked by providing or receiving at least one type of informal product or service. By the early 2000s, such arrangements were less common. Thus one resident of a Northern Peninsula village, when asked if people in the community helped out like they used to, responded as follows:

No, this is a hick town for that. If you can't depend on yourself you can't depend on anybody else around. Years ago, it was you used to be able to depend on people. Before my husband got sick, everybody used to run to him for help. But when he got sick I had no one to depend on.

It is not so bad for all, or in every place; however it is clear that outmigration and changing demographics are threatening the complex inter-household networks which were so crucial to maintaining communities. To take the simple example of shovelling snow out of the driveway: with young people gone, this becomes hugely burdensome for old people and can make such matters as grocery shopping impossible, which in turn can reduce food security.

CONCLUSION

On both coasts, and in both Indigenous and non-Indigenous cultures, the informal economy has been part of a society that built its economy on the extraction of natural resources, but one in which large-scale industrial production has rendered that economy vulnerable (Ommer and team, in press). Coastal communities were (and many still are) kin-based, and have functioned most effectively at the level of band, or extended family, seeking production for survival, not for profit in their informal sector (Brookfield 1972).[2] Informal economic arrangements and an egalitarian ethic have also been the norm (Brox 1972; Ommer 1991; Thornton 1979; Mannion 1977; Brody 2000): a kind of "moral economy" (Cadigan 1999). What has not been understood, but is of crucial importance even today, is that the formal economy relies upon the informal system. Historically, that was because the flexibility of the community and its seasonal exploitation of a range of resources allowed the local residence of an otherwise too expensive labour force (Ommer 1990b; Ommer and Sinclair 1999). For settlers too, diversifying resource exploitation was a means of survival. In more recent years, coastal communities have struggled to maintain what Sinclair (1985) has termed domestic commodity production[3] in the teeth of licensing policies that have made it impossible to move flexibly from one resource species to another as availability and prices would dictate, as well as the scarcity of cod inshore in many of the last thirty years. The flexibility of the earlier coastal system has been virtually destroyed by resource policies in the recent past.

The informal economy has also been transformed as coastal people have become more dependent on cash income from some source to combine with subsistence production. In this sense, the informal economy relies on the formal economy for the "start-up capital" that has made informal activities functional: the cash with which to mend homes, build fences, hunt for and harvest country food. A combination of wages, cooperation among households, and considerable subsistence production has allowed coastal residents a modest living supplemented, where need be, with employment for a while in another sector (or region). However, this functional, flexible, and highly adaptive system, which underlies the

---

[2] But see, for example, Brookfield 1972, pp. 9-17 for a close argument about the value of peasant societies and the possibility of their integration into colonial (and post-colonial) societies in a positive, rather than destructive, manner.

[3] The term refers to production of goods for sale based on household ownership of the means of production and the utilization of household labour.

Newfoundland boast (referred to in our introduction) that, even in the Great Depression, one would "never starve here," is now seriously threatened. As firms and government services have restructured, they have increased their own flexibilities but removed them at the grass-roots level of small communities. Multi-tasking and multi-skilling—once so much part of the informal economy on both coasts—has become the prerogative of industry, leaving local wage-dependent employees without the time and seasonal flexibility that used to sustain the informal economy. In Newfoundland the 1986 Royal Commission on Employment and Unemployment argued that the informal sector was an important sustainer of social cohesion, even in the face of economic collapse (Newfoundland and Labrador 1986). Recent research carried out on the east coast of Canada in the wake of the groundfish moratorium, has confirmed the Commission's argument, finding that the informal economy (legitimate economic support networks of kin and community) worked effectively to keep people productive even when they were not "employed" in a wage-earning sense. Indeed, these same networks were found to be of significant help in dealing with the stress that surrounded the moratorium that was suffered by rural communities: those who were operating within robust informal structures, it was found, had better mental health than those who were restricted to relying on the formal economic structures (Ommer and Turner 2004, 152).

However, there are limits in the extent to which the informal sector can take up the slack and facilitate household and community "adjustment" in the face of massive economic and ecological restructuring. Global restructuring is fundamentally affecting the capacity of coastal societies in Canada to exist, as the transition from a small scale local to an industrial society extends globally. An important implication of globalization is the threat to the flexibility that enabled households and communities to survive on the periphery.

While not romanticizing life in the informal economy, careful consideration needs to be given to preserving what is positive about it, as policy-makers and communities struggle to find a place in a rapidly changing global economy. It is possible that the world is changing too fast, not giving people time to adjust as impatient governments and industries seek to hold the national place in the global rush for competitiveness. It is also possible that such a rush is foolhardy, for its consequences, both human and environmental—while not yet clearly understood—are beginning to appear to be destructive of biodiversity and human culture. The rate at which things change is important: given time, small communities can adapt—indeed, their culture has had adaptive flexibility as a hallmark.

It is also the case that we need small communities. Their great strength has been their intimate relationship with the environments that have sustained their cultural life and their physical and nutritional well-being: their social-ecological health. They are the miner's canary in the social-ecological systems of Canada's coasts. When things go wrong at the local level, whether that be socially or ecologically, or (as so often) both, they are there to feel the disturbance and warn us. Thus, for example, inshore fishers, seeing smaller and fewer fish inshore were warning about collapsing stocks long (twenty years) before government officials finally recognized the crisis, perhaps too late. Moreover, small-scale ecological economies are intimately concerned with important flexibilities which have in the past, and could still now, enhance community resilience and provide a base from which ways forward could be developed. Seasonality, for example, has been and could be again, turned into a strength in a pluralist economy which will have to develop ecotourism, small-scale aquaculture, and perhaps "slow food." As the developed world is now having to recognize the wisdom of ecosystem resource management, such ecologically sensitive economic behaviour will be of use to us. Coastal communities have things to teach us, in a world which relies too blindly on technological fixes. The rural informal sector is a vital part of any future potential means of livelihood for Canada's coastal communities. It sustains them culturally, socially, and nutritionally, and we will be foolish if we lose the ecological wisdom that has sustained such places for centuries and even millennia.

In: Resetting the Kitchen Table   ISBN 1-60021-236-0
Editors: C. C. Parrish et al., pp. 129-144   © 2008 Nova Science Publishers, Inc.

Chapter 9

FOOD SECURITY, LEARNING AND CULTURE: BRIDGING THE RESPONSIBILITY GAP IN PUBLIC SCHOOLS

Carol E. Harris and Colleen Shepherd

Educational Psychology and Leadership Studies, University of Victoria, Canada

ABSTRACT

In addressing food provision in schools, we trace theories of technical rationality that focus on ever-increasingly purposeful models of thought and action, and neo-liberal ideologies that measure purpose solely in terms of efficiency. Largely missing from the technical standpoint are normative considerations of societal responsibility. Our findings, from an in-depth study of one high-risk BC elementary school, indicate that conditions of child safety and health have improved greatly since the introduction of food programs; that many children and parents choose their school because of its food program; that the school has implemented comprehensive provision that, in the view of school personnel, avoids stigmatization; and, finally, that there is room for improvement in the links between First Nations cultures and food provision. We question the "manipulated anxiety" that continues to threaten program sustainability, and the "burden of proof" that currently surrounds such programs. We note that administrators and teachers who work in conditions where children receive nutritious meals and snacks, perceive any other arrangement as "criminal neglect," and an abrogation of their responsibility.

INTRODUCTION

Opposing theoretical positions surround the efficacy of food programs in schools. One advocated by the World Declaration on Nutrition, endorsed by Canada and acted on to varying degrees by provincial governments and volunteer organizations, holds that the provision of food to hungry children brings about dramatic improvements in their emotional well-being, social behavior, and ability to concentrate (World Food Summit 1996). These changes, in turn, according to proponents of food provision, result in improved academic performance. This literature centers on issues of "food security" and, as indicated in the World Food Summit definition (Turner et al., Chapter 1, this volume), implies a basic human responsibility and right.

An alternative position, put forward by several Atlantic Canada researchers, maintains that the provision of school meals (usually undertaken as a volunteer effort), while conceived of as "wonderful" programs, in fact contribute to increased dependency (Dayle and McIntyre 2003; Williams et al. 2003), stigmatization of recipients (McIntyre et al. 1999) and a growing bureaucratization of program delivery (McIntyre et al. 2001). An important consideration accompanying these problems, as articulated by Hay (2000), is that poorly planned and implemented programs—that is, implemented without benefit of rigorous evaluation—can become smokescreens masking the underlying issue of income insecurity and the dismantling of social services.

Our position on meal programs at Roosevelt Park Community School in Prince Rupert, for reasons that will emerge below, supports the rationale of the World Food Summit. Hunger is seen within this and other coastal community schools as an urgent problem calling for immediate action. At the same time, we support the warning of Hay and others (e.g., Keating and Hertzman 1999) not to lose sight of the network of social and economic issues that underlies Canada's present unwillingness to redistribute wealth and stem the growing gap between rich and poor (Goudzwaard and DeLange 1995; O'Connor 1998; Townson 1999). In addressing the effect of restructuring on the health and well-being of coastal populations—the main focus of the research initiative, *Coasts under Stress*—we examine the perceived need for, and effect of, school meal programs.

THEORETICAL UNDERPINNINGS

To understand debates around school meals, and those surrounding the restructuring of education in general, we began our study with an examination of the larger scene. Accepting assessments of the western world's (and now, global) drift towards ever-increasing technical thought and rational ity (Habermas 1971; Heidegger 1977; Weber 1978), we place the role of such rationality in the present context of globalization. From the extensive literature on this topic, we point to several characteristics which, in a roundabout manner, reflect upon children's food security. One of the effects of globalization, with its priority on the market interests of large corporations and trade cartels, has been to lessen the power of states to determine their own affairs (e.g., Atasoy and Carroll 2003; George 1999; Ife 2002).[1] The ideology of globalization, with its inverted power between state and financial institutions, harkens back to liberal ideas of "laissez-faire," or market-driven economics.

The link between globalization (or neo-liberalism) within Canadian schools, governed as they are by the provinces, can be seen in the move to decentralize control and responsibility— at least ostensibly—from central ministries of Education to individual districts and schools and the volunteer community at large. The most effective means of purported decentralization comes about as individual districts, and thence schools, become responsible and accountable for the allocation of money for resources and capital expenditures. In contradiction to the

---

[1] Some argue convincingly that globalization has paved the way for states to relinquish control (Gregg 2005; McBride and Shields 1997).

rhetoric of decentralization, however, schools in Canada and elsewhere experience a tightening of centralized control through standardized tests, prescribed curricula, reduced budgets in operational dollars, and a variety of accountability mechanisms (Blackmore 1996; Foster 2004; Goodman 1995; Harrison and Katchur 1999; Smyth 1993; Taylor et al. 1997).

Contradictions and debates about educational processes and purposes have deep historical roots. Callahan, in his classic text on *Education and the Cult of Efficiency* (1962), described the entrenched market model of school economics in the early years of the twentieth century with its emphasis on scientism and the limited technical goals of efficiency and effectiveness, as recorded in standardized curricula and testing. These quantitative measures of outcomes were later challenged by John Dewey who drew attention to the importance of process and the active involvement of students in their own learning. The progressive approaches of Dewey and others were reinforced by the post-WWII focus on equality of educational opportunity and socio-economic equity. As early as 1957, however, the successes of the USSR in science and technology (i.e., launching in space the first satellite) signaled in North America a return to more competitive and conservative educational approaches. By the 1970s, with the additional downturn in economic growth, education once more was "under siege" (Giroux and Aronowitz 1985) as budgets were trimmed, and standardization deepened. Despite an emerging interest today among educators in social justice issues (EAQ 2004), accompanied by warnings about the limitations of data-driven rationality (e.g., Bromley and Apple 1998; Bowers 1988), the trend to quantify and market success continues and the pace accelerates.

Throughout this period, we have two major forms of rationality and action at work, technical and values-based.[2] The underlying rationality of technical (at times called "bureaucratic" or "instrumental") thought and action focuses on achieving maximum efficiency and effectiveness in reaching pre-determined goals. Within an economy of high capitalism, efficiency is frequently calculated in terms of doing each job with minimum monetary expenditure, while effectiveness relates to getting the job done well. Of course, the goals of saving money and effective action may collide. In political and economic interactions, we can observe many examples of this, one tragic example being the recent debacle played out at Walkerton, Ontario. In the late 1990s, hundreds of Walkerton citizens became ill as *E. coli* and other bacteria from manure spreading seeped into the water supply, resulting in seven deaths and many long-term health consequences. The deregulation of water testing as effected by the Harris government, in this case, became too heavy a long-term price to pay for short-term "efficiencies."[3]

CONCEPTUAL CONSIDERATIONS

Our approach to the issue of food security is based on three methodological assumptions. First, we hold that the complexity of human interaction calls for more than an assessment of

---

[2] See Brubaker (1984) for a particularly cogent discussion of Max Weber on Zweckrationalität (technical rationality), Wertrationalität (values-based rationality) and other forms of rational thought and action.

[3] Equally serious conflicts can occur as technologically rational purposes collide with values rooted in religion, philosophy and personal experience. From the standpoint of Liberation and Social Gospel theory, for instance, effectiveness is found in social action rather than economic bottom-lines (Crossan 1996; Freire 1998; Margoshes 1999).

goals and outcomes. We also look to the significance of personal and collective responsibility, as empathic "sensitivity to the weal and woe of others, a caring . . . that can both prompt beneficence and stave off harm and abuse" (Hatab 2002, 249), and as "being with others"—not of identifying with them or reducing their responsibility for their own lives but, rather, in walking with others in mutual benefit to one's self and persons in need. In this latter sense, according to Heidegger, a sense of responsibility is fundamental to human thought and action (Raffoul 2002).

Second, we believe that the best way to understand an issue—in this case, the impact of restructuring on food provision and children's health—lies in listening to the experiences of people directly involved in the issue. Thus we situate our study in a single school where we can focus on the first-hand accounts of many people.

Third, and consistent with participatory research methods advanced by *Coasts Under Stress* (e.g., Bannister 2003; Harris 2004), we hold that those working with communities, in either the social or natural sciences, need to recognize and draw on local and traditional knowledge of community members throughout all stages of an investigation—that is, while planning, gathering information, asking for feedback, and formulating reports. From the standpoint of university-community relations, we pose the following questions: What importance do men and women, involved in teaching and nurturing children who are widely recognized to be "in need," place on school meals and other food provision? What have been their experiences of food provision and its absence?

DESIGN AND METHODS

For our detailed study of food security as a significant aspect of restructuring, we selected in 2000 a "community school" in Prince Rupert, BC. Prince Rupert is a coastal city buffeted by collapses in its fishery, forestry, and transportation industries, and a population decrease over five years of 6,000 (from 17,000 to 11,000 between 1995 and 2000). When we began our study, eighty-six schools in the province were designated as community schools because of their high-needs populations. Roosevelt Park Community School, with grades kindergarten to seven, serves an 80% First Nations (and off-reserve) population, approximately 10% Asian immigrant families, and 10% Caucasian families, all located in or near the catchment of an inner-city housing estate. In step with the city de-population, Roosevelt's student numbers decreased from 350 to 250 during our study period (2001–04). This school "on the brow," despite problems brought about by ever-deepening poverty, has gained a reputation for its safe, caring environment and its focus on student literacy (Harris et al., forthcoming).

We first focused on beliefs, attitudes, and values as expressed by school and community participants, interpreting these in light of the rationalities discussed above. The data, collected in six visits to Roosevelt, came from two main sources. The first involved a series of in-depth interviews with the principal, Steve Riley,[4] and thirty other interviews with school associates, including support staff, teachers, community outreach workers, district personnel, and

---

[4] We are deeply indebted to Principal Steve Riley who met with us many times, read our work and discussed with us our developing ideas about food services. In many ways, he epitomized the "inside researcher" who became "a powerful lever for personal, professional, and organizational transformation" (Anderson and Jones 2000: 428).

parents. Participant/observations provided the second data source as we, the authors, took part in the administrative, social, and pedagogical life of the school.

We begin our report with a brief history of Roosevelt Park and how it came to receive funding as a community school. Next, we provide an overview of food security prior to government funding. This is followed by our findings concerning the present food programs. In our discussion, we re-visit the objections to food programs raised by other scholars, and add perspectives obtained from this qualitative research.

COMMUNITY SCHOOL PROGRAMS

From the early 1990s, Roosevelt and two other city schools of Prince Rupert applied successfully for funding to supplement special programs. The first grant, according to a former Roosevelt principal, enabled the schools to offer lunches and, soon after that, breakfasts. The next initiative for Roosevelt was to seek and obtain an Inner City School grant which brought funding for an "opportunities program," with an additional teacher to develop special interventions for high-risk children, and "help the staff understand some of the underlying social issues facing these children." The third initiative was to apply for community school status. By this time, Roosevelt had demonstrated exceptional need, and its staff had launched several successful enrichment programs. The new designation, when achieved, brought with it an annual infusion of $75,000 to be used for additional teacher aids, greater program enrichment, and community outreach. This major government allocation, supplemented by special grants from service organizations, has supported initiatives in student and adult literacy, summer and evening educational programs, special classes for children suffering emotional trauma, and a host of school enrichment activities.

Such targeted school funding is now a thing of the past; since 2001, the provincial government has allocated money directly to school districts, according to enrollments and statistically arrived at measures of population need (i.e., records of employment, social assistance, and so on). While funding still comes from two government sources—the provincial Ministries of Education and of Children and Family Development (MCFD)—that emanating from Education to the District was reduced in 2003 by $200,000,[5] while infrastructure and funding from the MCFD had largely disappeared.

Because the District fully realizes the special needs of its inner city schools, Roosevelt was in June 2004 still maintaining most "essential" services—additional childcare workers, a teacher to work with children "at extreme risk" of academic failure or emotional collapse, two community outreach workers, some special evening and summer programs, and classes for guardians, parents, and kindergarten children to practise literacy skills together. The school continues to serve breakfasts to children, lunches to all, and a fresh fruit snack at the end of

---

[5] This figure appears in the financial report of the District for the school year ending June 2004. While the Ministry of Education claims that money has not been cut from education budgets, this fails to include calculations of increased teachers' salaries. The new formula for funding falls under a plan called CommunityLINK funding, the acronym standing for Learning Includes Nutrition and Knowledge
<http://www.bced.gov.bc.ca/communitylink/>

the day. Nevertheless, each year new funding cuts threaten the food quality and the level of program enrichment.[6]

FINDINGS

Before Formal Food Programs

In order to form a sense of conditions before food programs were introduced, we met with teachers who had been part of the school for ten years or longer. When asked what it was like then, all teachers agreed that hunger was much in evidence, and that it affected the eagerness and ability of many children to concentrate on their work. Marilyn,[7] recently retired from teaching, remembers that "those teachers who provided food often fed only their own classes." She had been doing it for years when she discovered that other teachers, too, did "not want hungry children in [their] classrooms . . . Then when we found that so many of us were feeding our students, we put the sandwiches out in the hall and did it on a broader school level."

In those days, only the hungry children took the food or, as Marilyn notes, the "brave kids who weren't too embarrassed to ask for food." Marilyn's ploy, to avoid the stigma noted by researchers McIntyre *et al.* (1999), was to prepare a table where all the kids would share their food. "Nevertheless," she recalls, "we didn't get them all, not by a long shot."

Nor were all teachers in agreement about food provision initially. Some of them felt that this would lead to an endless chain of dependence on "hand-outs." Whether or not that would have been the case, Marilyn contends that, today, teachers and support staff are relieved, "just to know that the children are not hungry. You used to see children, those you had sent out at lunchtime, and they would just sit in the park or under the trees. They would never go home, [because] there was nothing to eat there."[8] Marilyn remembers that, once the food programs began, "You saw a happier lot of children all around you, more cooperative in the halls, on the playgrounds. The atmosphere was altogether better."

When we asked Dianna, a First Nations teacher, about former times, she spoke of conditions that differed according to the season and time of month. She recalled, for instance, distinct changes towards the end of each month, as money supplies dwindled. Then she saw "a lot of fighting in the classroom, kids not working and sometimes coming in really hungry [because] they just had too much on their minds." She could not "stress enough" the importance of the lunch program: "It's probably the backbone of what is happening at Roosevelt. My kids went to the school before the program. The lunch program made the major difference, because you can't learn when you're hungry." This teacher, like Marilyn,

---

[6] At the time of writing, a provincial election date looms (i.e., May 17, 2005). This date is accompanied, contrary to the cuts recorded in the years of our study, by a sizable outpouring of government funds for social services. Because of this fluctuation in funding, we find it difficult to differentiate between real and anticipated losses in school food services.

[7] Because of the sensitive nature of participants' testimony, we use pseudonyms for teachers, parents, and social workers.

[8] Several participants offered two versions of this story. One version described children staying near the school after hours, knowing they had no food at home; the other centered on the hording of food from school lunches in order to feed hungry family members.

assessed the differences in better social skills among the children and an improved school ambience.

Riley, who worked previously as a vice-principal in one of the city high schools, noted there "unequal conditions, where some students have food and others do not." He does not want Roosevelt students "going on to the high schools, sitting next to middle and upper class kids who have lunches every day." Riley knows the hungry ones will become alienated and "disruptive. They'll make those kids' lives miserable. Basically [without food programs] you are saying society doesn't give a damn. Kids are really quick to figure that out."

For about a decade now, food has been part of the Roosevelt child's day. The school secretary, according to Riley, is the first to spot those who arrive at school hungry. Lana sees that they "get a basic breakfast of cereal and milk." She assures us that the food programs are a "very important consideration for many parents as they enlist their children in the school. You can almost guarantee that someone is going to ask the initial question, 'Do you have a lunch program?'"

Food Programs as a Survival Issue

This question about provision, posed by many of the families who circulate in and out of the city schools,[9] attests to the importance of food security to recipients. From the perspective of community outreach workers, as well, the programs are essential. Tony, a First Nations outreach worker, deals with children at the most needy end of the spectrum. He notes that "students come here some days not to learn, but because they have issues. You can't expect them to learn. They sit there and they will have a blank face. You know there is something really going on." Tony cannot separate the food issue from the many other problems facing these children. For him, "there are just too many things that come into it. I mean if food's an issue, then there are other things that are issues. If they can't afford clothing, if they can't afford food, it just becomes a whole survival issue, a day-to-day thing and a struggle."

Shepherd, who conducted this interview, remarked that Tony must have a tough job, trying to fit together the complexities of his work—"all the different pieces." "Yes," Tony concedes, and

> when they start to come together, it gets frustrating. [For example] I'm sitting here with a social worker who told a parent, "Look, if you're short or you don't have enough food, go to the Salvation Army. Ask for [name of person] and tell them I sent you; they'll give you enough to get you by." You know, to have that happen, you could just see the pride. Over the past few days, I've been thinking about these single dads that are raising their kids. And [the problem] seems to be growing; it's a survival skill that's happening because of tight times.

In an era in which food banks represent a "survival skill," school food programs become ever increasingly significant. As Steve points out, "You simply can't expect kids to develop when they are hungry, if they're not loved, if they're not fed, if they're not sheltered." In this school that prides itself, despite its socio-economic setting, on remarkable literacy gains (Harris et al. forthcoming), Steve asks poignant questions: "What relevance has learning to

---

[9] We were told that several families in the area tend to be evicted from one housing estate after another. At the end of the month, they leave one "home," only to return when funds run out at their new location.

read if you watched mom get beat up last night, or if there's no food in the house again and the baby was crying all night?" He adds that "these are realities not for one kid in here, not two, but for hundreds."

Anxiety over Real and Anticipated Change

Much of the talk in the school centered round the changes people were experiencing, and those they feared. Although teachers and support staff at Roosevelt continue to provide the extra amenities needed by their student clientele, teachers identify areas in which monetary cuts are affecting the school. First, class sizes are creeping up, although Roosevelt can still afford the child care workers who help individual students. Second, fewer funds mean fewer evening and summer programs, for children and parents alike. Third, a reduced meal budget militates against the little extra celebrations, such as pizza day. Perhaps the most disturbing feature of the cuts is the sense of uncertainty experienced by school personnel. Rumor and generalized announcements from the government of future "belt-tightening" combine to produce tension, anxiety, and fear. As school problems multiply, school and community workers are conscious of the larger socio-economic scene. One outreach worker put it this way: "You're starting to see families breaking up, kids coming into [foster] care and some parents who say, 'Look, we don't have the funds, we can't buy food.' And when a parent has to tell you that, you know it's bad. And it's happening more and more now."

Not surprisingly, the major fear was of job losses combined with deepening government cuts to assistance. Even now, according to Tony,

> Houses are getting crowded because the social assistance is cut and there are no jobs in order to pay the rent. You get two families coming together and then you have chaos—and we're just starting to see this happen. Then I think of the timetable for people here to be totally cut off, completely, and when we hit that, we're in trouble. I don't know how they're going to manage it.

That the government relented on the issue of total cuts to welfare recipients, mentioned in this 2003 interview, does not diminish the concern, worry and fear felt by unemployed men and women and their families. It is reasonable to expect these anxieties, in turn, to be experienced by the children of affected families. They certainly appear to be internalized by the administrators, teachers and support staff of Roosevelt.

Another anxiety surrounds the dismantling of existing programs, and the establishment of new program designs. At present, food programs are managed by employed workers. Principal Riley, who has felt pressured by District administrators to find volunteer replacements for these tasks, points out disadvantages in the volunteer system. These include additional supervisory work and higher costs. "In order to get volunteers in the school to make 220 lunches a days," he knows he "will have to get more vice principal time." Then, too, "the cost will rise—not out of the meals program, but because of administration costs." At present, one woman does the books, the ordering of food and "all the rest of this major production." As for volunteers, they "don't always show up. What do you do when your volunteer gets sick or falls and breaks her leg? You know their kids get sick; [sometimes] and they can't make it in. What do you do? You have hungry kids in the school; you don't feed them that day?" Riley's present practice of hiring a worker—as opposed to buying the food from an outside source—saves, he estimates, $40,000 dollars annually from the school budget.

From a humanitarian point of view, Riley considers the volunteer route merely an extension of the food banks which, in his experience, "truly stigmatize recipients." Although his church has a soup kitchen very near the school, he does not "go near it, because it's embarrassing for the kids and parents of this school." In this regard, Riley notes the strange logic of a country that can afford armaments but "cannot afford to feed its poor."

Linking Nutrition and Culture

Food workers at the school are aware of two important issues pertaining to their programs; one concerns nutrition and the other, indigenous culture. So far they feel fairly confident in the nutritional quality of food. Mary, the food planner, works "with a nutritionist, a community health nurse actually who has looked over some of our menus." Mary believes the quality is quite good and that the menus are interesting. She recognizes, however, that "it's getting harder to provide good food when the budget gets cut. We have to make more compromises you know—we have to slim down our menu options and purchase foods that cost less." "Nevertheless," she adds, "we always offer healthy snacks like fruit or vegetables. More and more the kids are choosing good food when it's available." We were able to observe students lining up after classes each day for a special treat—usually an apple or banana or something equally beneficial to their nutritional needs.

The cultural aspect of food provision takes on special significance in the First Nations context of Roosevelt. While the meals and snacks overwhelmingly reflect food choices of the dominant society, First Nations teachers and support workers try to involve the children in special food celebrations of their own culture. Tony, for instance, recently invited several of the First Nations boys to attend a ceremonial feast of salmon. A community Elder with whom we spoke reinforced the importance of trips like this and the learning they generate, especially in the context of cultural loss:

> My people were proud, we were free and healthy . . . now look at us. Our children are hungry and our fathers are angry. We don't have jobs and we don't have the fishing and the food from nature the way we used to. Now that we moved to cities, and have changed our ways to fit the white ways, we are lost. Food is the thing that keeps us all together you know, it still is the thing. We feast like no other people. Our feasts are where the Elders teach us and teach the children and the fathers and the women. I learned. I learned so much from the Elders when I was a child. My learning came with the Earth and from the earth and the oceans and went into my mouth and fed my spirit not just my body. My teachings become part of me and will go back to the earth when I am gone. I remember those days.

> I am sad that we are poor; we have always been rich. I don't mean with money, I mean with food. Even if I had no money I could always eat. Now we go hungry and we have no place to teach our children. We still teach them but it is not the same. It does not feel the same. Why can some people eat but some people have to go to the soup kitchen or go hungry? Why is it like this? I believe that all people should have the same. It all belongs to the earth. It is not mine to own, it is not yours to own. That is all.

The Elder's words, which we record in detail, indicate not only the significance of food rituals to his people, but also suggest directions in which the school can move as it plans future food programs.

Some Things Just Can't be Proven!

We asked each interviewee about the contribution, overall, of food programs to the school and its children. In the responses, we detected more than a little frustration that the question, in the context of Roosevelt's obvious need, was even posed. Mary, a support worker, exemplified this attitude, as she urged us to "just ask the teachers around here."

> I see it all the time—kids who eat, learn better. That's common sense, of course, but I've witnessed it myself. The behavior problems balance themselves out as kids' bodies become stable and healthier. There are kids that I know who were sick from malnourishment and who exhibited, you know, behavioral and thinking troubles. After regular feeding, those kids have really bounced back. But you know you can't prove it directly, I mean, the link to school meals. Some people say well maybe the kid is getting other things outside of school—so you can't prove it is about meals in school. Well, I say good; I hope they are getting something outside of school, but I'm telling you that regular school meals have a huge role to play.

This conviction about need was reinforced by everyone we spoke with. Jonathan, a social worker who deals with the full spectrum of challenges in the area—poverty, violence, hunger—had this to say about the children of Roosevelt, many of whom come from his client families:

> I know their lives intimately and I know the real concrete value of the food programs to those children. I have seen great improvements in kids who have moved into the [Roosevelt] catchment from a baseline of ill health (when they were at a school with no lunches) to a marked increase in health. Many of my clients are rather transient, but when they get to this school they try to have their kids stay there long-term due to the breakfasts and lunches. So I see how kids perform better, behave better and look better when they started eating regular breakfasts and lunches.

Obviously deeply involved with the families in this part of town, Jonathan acknowledges a responsibility for their welfare, as well as his frustration that this is not more widely shouldered:

> It would break my heart if they lost the food; the kids would go downhill for sure. My question to [those who make decisions about funding] is: what is the ethical responsibility to those kids, and who is to be held responsible? Those kids go to school for five to six hours a day, five days a week, eight months of the year, twelve years of their life. Who then, during all that time is responsible for the health of those kids?

Gordon, another government worker, while not pretending to speak from "a medical perspective," has this to say:

> I can't physically measure the kids before and after health [initiatives]. But what I can do is tell you from experiences and observations. I believe that the matrix of health-related issues

from energy and vitality, to cognition and comprehension, to memory, to behavior, to social well-being to mental health, are all affected by nutritional levels.

Like Mary and Jonathan, Gordon rejects the very suggestion that "school meals do not enhance the nutritional health of kids" and the "argument that school food may be unhealthy. First, some food is better than no food at all, let's start there. Then, let's advance a little further to consider that most meal programs attempt to provide a minimum balance of nutrition." Gordon, who has worked on the coast for more than a decade, considers school meals to lie at the core of security for these children, many of whom "have no [home] safety net whatsoever."

Gordon recognizes that our entire conversation about food must be viewed in the context of other collapsing social programs. In his department, where staff numbers are decreasing rapidly and workloads are increasing proportionately, he sees that caseworkers are unable to fill the social needs of families. In many instances, "kids and families are falling through the cracks." In a time when "someone needs to care," he sees Roosevelt—and especially its food programs—as providing this service.

DISCUSSION

This theme of "falling through the cracks" of social services permeated our conversations at Roosevelt. Everyone was conscious that cracks are inevitable in the present system of rapid economic and bureaucratically mandated social change. During a single visit to Roosevelt in 2004, for example, an at-risk teen girl was murdered, and the mother of children at the school died of a drug overdose. These tragedies, grieved by school personnel and considered preventable by many of our participants, brought home to us the reality of deteriorating conditions in this coastal city. When faced with personal problems of this magnitude, the arguments of those who see "truth" in terms of numerical data alone appear particularly inadequate. From another perspective, we begin our discussion of this school with a focus on two major findings from our study of food programs; these we term "manipulated anxiety" and "cultural cohesion."

Manipulated Anxiety

Our participants, in agreement with Hay (2000, 10, 11), identify family income insecurity (that is, poverty) as underlying child hunger and accompanying ill heath, stress, family violence, and even illiteracy. It became obvious to us that the social safety net has been severed in many places (Riches 1986; 2002) by two main players. First, the federal government, while signing international agreements to lessen child hunger and poverty, between 1993 and the late 1990s cut transfer payments to provinces for health, education, and social services by approximately 20% per annum. Second, policies of restraint in social spending, initiated in Ottawa, picked up scope and momentum under the neo-liberal ideologies of the Ontario, Alberta, and BC provincial governments. Choices made in the late 1990s and early years of this decade, particularly in these provinces, led to tax breaks at the expense of social programs and to a greatly widened gap between rich and poor citizens (Kerstetter 2003; O'Connor 1998).

We see that current practices, at both federal and provincial levels, pose the dominant threat to educational and health security. The provincial government of BC, with an imminent election looming (in February 2005) and advance polls indicating a tight electoral race, announced many new funding initiatives. One, pertinent to this study, promised an infusion of $150 million targeted for school libraries, special education, and music programs—the very programs slashed by indebted school districts during the previous four years (*Times Colonist*, 1 February 2005).[10] Not surprisingly, the impact of these political practices for food programs—promises of spending in the months leading up to an election and severe restraint at other times—produces a continuous state of anxiety among administrators, teachers, parents, and children. While Roosevelt experienced few large cuts during our study period— apart from those observable in reduced number of teacher-aides, larger class sizes, and meals of lessened quality—school personnel, constantly reminded of imminent reductions, were unable to plan rationally for their outreach and school programs, as well as for food provision.

Cultural Cohesion

Our second major finding addresses program planning as well, but in a more positive vein. While food is crucial to physical health, and we have provided ample evidence that the food offered children at Roosevelt is nutritious, we suggest with Counihan and Van Esterik (1997) that it is equally important to recognize that the significance of food extends beyond bodily sustenance, into the very the heart of social relations and deeply rooted cultural meanings (Harper and Le Beau 2003). To eat and drink together is "at the same time a symbol and a confirmation of social community, and of the assumption of mutual obligations" (Freud, in Counihan 1999, 125). Behaviours and belief systems surrounding the production, distribution and consumption of food reveal a great deal about power relations in any society. In this respect, food becomes both a "product and mirror of the organization of society on the broadest and the most intimate levels" (Counihan 1999, 6).

Economically disadvantaged children, such as those in our study, are often socially isolated, unable to take part in the same cultural food activities as their cohort group (Riches 2000). Alienated in both the public and the private sphere, poorly fed children often experience a deep sense of social exclusion (Counihan and Van Esterik 1997). Not surprisingly, social stigma and maladjustment commonly propel these children into troubling futures. This possibility holds particular relevance for educators, as schools provide primary sites for the social experience of children.

The Elder, quoted in length in this chapter, reinforces in the context of BC the cultural importance of food sharing to his people. He indicates that important lessons, symbolized for him and his people in partaking of certain traditional foods, are passed down from generation to generation. For the First Nations people of BC, and the transient indigenous children of Roosevelt, such knowledge becomes an essential component of collective pride and cultural

---

[10] The Government website for the Ministry of Education on 6 February 2005, outlined other details of promised funding as $180,000 to sixty schools (unnamed) for "excellence " (announced January 17), $5 M for literacy innovations (January 27), and $5 M for a francophone school in Kelowna (2 February).

identity. First Nations participants in our study were particularly aware of the larger implications of food programming that might, in time, extend mere food provision to include culturally appropriate foods and celebrations.

Stigma and Dependency

We now move to the relevance of problems, identified by east coast researchers, to the context of Roosevelt. The first of these is stigmatization (McIntyre et al. 1999), a condition reported by our school participants prior to the introduction of food programs. In our study of present conditions, however, stigma emerged only as a possible by-product of cultural exclusion. Common stigmatizing situations, such as occur when children view themselves, and are viewed by others, as objects of charity, are largely avoided in BC. At Roosevelt, for example, all children (except those without need and in the immediate vicinity of the school) sign on for lunches and snacks. As the principal pointed out, only the Lunch Lady knows which families are unable to meet the daily cost of one dollar. Further, as programs are planned and run by paid workers, and not by volunteers, food programs avoid the stigma (and the administrative hurdles) of becoming "just another charity" like the city food banks, already frequented by growing numbers of the Roosevelt parent body.

While school food programs are viewed by providers and users alike as "wonderful," researchers contend that, in Atlantic Canada, they can contribute to increased dependency (Dayle and McIntyre 2003; Williams et al. 2003). Our evidence, from this one BC school, indicates that this danger, in the face of extreme need, must remain secondary to immediate action. An emphasis on dependency assumes a victim-blaming stance, all-too-often embedded within welfare/social policy reforms, that directs public attention from a model of mutual responsibility and caring to one of personal obligation (Klein and Long 2003). With this shift comes the assumption that the current system makes welfare and other helping services too attractive. Ultimately, the "myth of dependency" (see NFA 1997) shifts the debate away from issues of distributive justice and social equity, to focus on personal deficiencies of the economically disadvantaged.

The Burden of Proof

The final set of inter-related arguments that we address posit (1) inadequate proof that food programs in schools do what they are established to do, that is, alleviate hunger and provide nutrition, (2) that few measurements exist to substantiate anecdotal claims of need, and (3) that objectives over time become diffuse and self-perpetuating.

In a review of school-based food programs across Canada prepared for the Federal Government, Hay (2000) asserts that presently available evidence fails to demonstrate that school food "programs are a sound social policy response for [child hunger]" (p. 4). This finding appears to be based on two related conditions; the first points to insufficient measurements of "real" efficacy, and the second to the inadequacy of anecdotal evidence. Williams *et al.* (2003), for instance, contend that "the benefits of children's feeding programs are . . . spoken of with conviction and certainty, despite their subjectivity, speculativeness [sic], and the absence of evaluative information or, in the case of parents, personal knowledge for justification" (p.168). And Hay (2000) indicates that "school food programs have tended to be implemented in response to a fairly informal, and sometimes anecdotal, assessment of a school's or community's social and economic situation" (p. 2). These researchers caution that a lack of demonstrable research leaves the value of programs open to mere speculation and opinion.

Our findings challenge this search for numerical verification of something as obvious as hunger and its alleviation. While studies from the Atlantic provinces of Canada are based on a different population, it is hard to believe that any school is without needy children. We question this sole emphasis on numerical data, fearing that vulnerable children will become lost in the mystification of "facts." Embedded within this approach is the assumption that *perceptions* and *meanings* of child hunger and of the value of food programs to children as held by parents, teachers, support workers and school administrators are imprecise, uninformed, and easily influenced. As a result, these perceptions cannot be counted as real knowledge or as sources of data for evaluation. The participants of our study who recalled situations before and after food programs were introduced, and the case workers from the government ministries who spoke about school meals as one remaining "lifeline" for children—when all else was collapsing around them—present a compelling argument for immediate action

The theme of inadequate proof also involves a claim that, over time, feeding programs take on objectives above and beyond the original ones of alleviating hunger and providing nutrition. Williams *et al.* (2003), summarizing findings by Dayle and McIntyre, contend that once in place, programs "shift beyond their initial focus . . . [instead] they purport to improve children's learning, model good nutrition, and relieve family stress for a wide catchment of children, not only the poor" (Williams et al. 2003, 163). These putative objections to food programs, in the context of Atlantic Canada, appear to our participants as value-added features of food programming. The people of Roosevelt school, and of its larger community, consider hunger alleviation the first and physical priority, but one inextricably intertwined with children's greater attention to their lessons, improved behavior to others, and general expressions of happiness. Progressive policy and program evaluations, in many contexts, encompass a variety of indicators along physical, social, emotional, and ethical points of reference (Blackmore 1996; Coward et al. 2000; Riches 2002). Most importantly, successful policy evaluation, like the initial formulation and implementation of policy, includes the active participation—as well as the voices—of those who are being affected by the programs (Bannister 2003).

The final argument to be addressed here, and the one strongly supported by our study, is that school feeding programs often camouflage serious problems of income insecurity and the dismantlement of a social safety net (Hay 2000; McBride and Shields 1997). Poverty-related food insecurity, unfortunately, receives little more than token attention from governments and domestic policy makers who are gradually downloading their responsibilities to the voluntary sector in emergency relief programs such as food banks and charity-based school meal programs. The danger, illuminated by arguments of insufficient proof and misplaced objectives, is that socially conscious opposition to food programs too often plays into the hands of those who call for less government intervention, individual (as opposed to social) responsibility and voluntarism (Riches 2002).

SUMMARY CONCLUSION

We address food programming from a population health perspective which views food as more than a private concern or a mere economic commodity; we see it, rather, as foundational to the health and well-being of individuals, families and communities, and to social and cultural stability. Our findings reveal the importance of inter-connected solutions—such as one-on-one attention to students, and an emphasis on literacy, care, and food programs—to combat the inter-related societal absences of adequate shelter, clothing, emotional security, and food. We view positively programs that broaden their scope to include the use of food provision as a social stabilizing force, as a site of learning, and as a way to build a caring school environment for vulnerable children.

While school feeding programs cannot, and do not, alleviate the structural underpinnings of poverty-related child hunger, they address the most basic and immediate human need, and right, to eat and be nourished. From a moral and ethical perspective, we cannot theorize hunger away, nor can hungry children wait until we solve the structural problems of our time. The population profile of Roosevelt school convinces us that its children are experiencing acute levels of social and economic exclusion, rendering them among the most vulnerable in Canada for poverty-related food insecurity (Klein and Long 2003). For our participants, food provision within the school is considered together with expectations of academic growth, a spirit of safety and trust, and a network of special services. The latter include counseling for individual students, outreach to adults of the community, and programs for students especially hurt by poverty, social disruption, and family violence.

Our findings at Roosevelt Park Community School lead us to offer educators and policy makers several recommendations for action: greater attention to the voices of people affected by food programming, the continuation of school food provision as providing by salaried—as opposed to volunteer—workers, the incorporation of culturally relevant foods and food celebrations, and the kind of stable funding from governments that allows for rational planning at district and school levels.

Although many of our findings do not corroborate those of researchers in Atlantic Canada, we must give serious consideration to the problems they identify: stigmatization, dependency, misplaced objectives, and inadequate proof of need. We ask ourselves about the different contexts of east and west, and how findings can differ so greatly. One difference may lie in the greater extent of traditionalism and rural living found in Atlantic Canada, and people's willingness there to help one another. Another may be attributed to our different methodologies: one study was national in scope, others involved a qualitative analysis of nine cases across four Atlantic provinces, whereas our study focused on a single school in considerable depth. The BC school, as well, had a high First Nations population while the school populations in the Atlantic programs were described only as mixture of urban, suburban, and rural.

The distribution of wealth by region across Canada, however, reveals another interesting comparison. While the gap between rich and poor in Canada as a nation is immense, and has continued to widen since the 1970s, the distribution in 1999—that is, the share of wealth held by the lowest five deciles of the population as compared with that held by the highest five—shows much less disparity in Atlantic Canada than in BC. Moreover, roughly 22% of family units in the Atlantic regions fall in each of the two middle wealth groups, compared to only

12% in BC (Kerstetter 2003), showing a marked skewing of wealth in BC to the upper and lower income levels.

Hertzman and Siddiqi (2000) throw light on the implication of this distribution for both coasts. They note, albeit on an international scale, a positive correlation between swift economic/political transformations and economic decline, as well as a growing distrust of civil institutions in rapidly changing societies. These authors also point to what they call a "virtuous cycle" of trust and prosperity in countries undergoing an increased redistribution of income. Unfortunately, Roosevelt's community and province, and Canada as a nation, illustrate the first scenario. The rapidity of restructuring has left this northern town, despite its natural resources, in a state of economic and social crisis. While some people have maintained their jobs and professional positions, many others have sunk increasingly into abject poverty. In the province, a similar pattern is emerging (Fuller et al. 2003, 5) as health and educational costs shift from the public purse to the private and individual sector. The policy implication for BC, in concert with recommendations of Hay, McIntyre *et al.*, is to effect a reversal of economic policy, federally and provincially, towards a more equitable redistribution of income and wealth.

In this study, we followed the lead of those within the school who claimed that "poverty and hunger are massive problems in the town." We spoke with health workers, social workers, town administrators, parents and business people—always with the same story of a flourishing economy that had collapsed, affecting everyone but leaving the poor in an extremely vulnerable position. We contend that further economic policies of severe restraint, if imposed upon desperately needed social programs, will result in short-term economic savings and long-term problems of social unrest. Adequate meals, alone, do not correct wider societal problems but, in the views of the people we met, they provide a first and necessary stepping-stone.

In: Resetting the Kitchen Table
Editors: C. C. Parrish et al., pp. 145-159

Chapter 10

CELEBRATING AND CONNECTING WITH FOOD

Susan Tirone[1], Blythe Shepard[2], Nancy J. Turner[3], Lois Jackson[1], Anne Marshall[2] and Catherine Donovan[4]

[1]School of Health and Human Performance, Dalhousie University, Canada
[2]Educational Psychology and Leadership Studies, University of Victoria, Canada
[3]Environmental Sciences, University of Victoria, Canada
[4]Community Health and Humanities, Memorial University of Newfoundland, Canada

ABSTRACT

Celebrations are an umbrella under which people commemorate their lives. Celebrations may also serve to honor a living person or group, or serve as occasions for communal work, which often includes feasting, or the sharing of food. Food events are some of the most powerful ways of forming a collective or group identity. This chapter is about the role of food in celebration and in developing one's sense of self as a community member. In this chapter, the focus is on the celebrations of three groups of people who live on the east and west coasts of Canada and whose lives have changed as a result of environmental, economic, and social restructuring. The way in which they commemorate special occasions reflects how people in our studies coped with the changes that are impacting their lives, how the cultures of British Columbia and Newfoundland and Labrador help shape the way in which people celebrate and form social connections, and the resilience of people who have adjusted their ways of celebrating to address new circumstances. All of the people we worked with or interviewed and the communities in which they live, have been affected by change due to major restructuring. The circumstances faced by our study participants are not unique; other Canadians and people throughout the world encounter change as well. Opportunities to celebrate with food provide people everywhere with the means for connecting on many levels. As change impacts the food available for celebrations, the occasions that people value, and the context in which they occur, further studies are required to explore how celebrations may change in changing times, and if important opportunities for connecting are being created, or lost.

INTRODUCTION

Celebrations are an umbrella under which people commemorate their lives. Throughout history celebrations have served to honour life events including births, deaths, weddings, moments of transition from one season to another or from one stage of life to another, or historical events. Celebrations may also serve to honor some living person or some group, or serve as occasions for communal work that often includes feasting or the sharing of food. Food events are some of the most powerful ways of forming a collective or group identity. This chapter is about the role of food in celebration and in developing one's sense of self as a community member. In this chapter, the focus is on the celebrations of three groups of people who live on the east and west coasts of Canada and whose lives have changed as a result of environmental, economic, and social restructuring. The way in which they commemorate special occasions reflects how people in our studies coped with the changes that are impacting their lives, how the cultures of British Columbia and Newfoundland and Labrador help shape the way in which people celebrate and form social connections, and the resilience of people who have adjusted their ways of celebrating to address new circumstances.

Celebrations are often thought of as happy occasions, possibly playful, but may be occasions for reverence and serious contemplation. For most people celebrating implies a social occasion and the presence of family and/or friends. Participation in the creation and orchestration of a celebration at times is a fulfillment of our duty and obligations to family, church, or community and contributes to the maintenance of social order (Huizinga 1971; Rojek 1995). For example, special occasions celebrated in a certain way, reinforce and confirm the roles of family and community members. Deviation from those roles may have serious consequences as Bella (1992) discussed with regard to Christmas. Bella noted that women tend to carry the majority of the load required to make Christmas happen for others. The work required to make Christmas and other special occasions happen may mean that women do not enjoy all aspects of the festive season, but celebration for others depends on their adherence to traditional roles as organizer, baker, gift buyer, decorator, and entertainer.

Celebrations involving food events are a shared sequence of emotional experiences based on symbolic interaction. Food celebrations are complex and important social phenomena that merit close inquiry and analysis. These events can bring a sense of consistency to life when denoting special times of the year. As well they are important because they instill a sense of belonging and togetherness, and a sense of respect for others. They are also opportunities for youth to explore and test the boundaries of acceptable behaviour among peers and within their communities and to learn about cultural norms and expectations from adults and leaders in a safe, supportive environment. Food celebrations provide important contexts in which youth can develop a sense of self through connections to others in the larger community. As youth move through new situations and contexts, they discover, attribute, and create meaning about themselves as social beings (Peavy 1992).

For the purposes of this discussion, the term celebration is used in the broadest of contexts. It is the activity people engage in to mark an occasion, such as the culmination of an important occasion as in a party following a sports competition. It can be as simple as a Friday night with friends, pizza and a movie, or as complex as a wedding, a community pageant, festival, or potlatch. It may be spontaneous, relaxing, and simple or rowdy, ostentatious, and carefully orchestrated. It invokes a "play-mood; one of rapture and enthusiasm and may be sacred or festive in accordance with the occasion" (Huizinga 1971, 10).

In this chapter we present three different stories of celebration in the lives of people who were the focus of *Coasts Under Stress* research in British Columbia and in Newfoundland and Labrador. First, is an account of an annual feast held in the Gitga'at community of Hartley Bay, on the northern coast of British Columbia, which for many years was hosted by Eagle Clan matriarch Lucille Clifton. The second piece provides an insight into youth in northern Newfoundland and how they celebrate special occasions ranging from simple Friday nights out with their peers, to family events designed to commemorate birthdays, funerals, and Christmas. The third is about a high school in a northern Vancouver Island community and how it has used food in an organized setting as a vehicle to create opportunities for youth to understand who they are and how they might be in the world.

FEASTING AT HARTLEY BAY[1]

On 21 June 2003, the Gitga'at community of Hartley Bay hosted a feast to celebrate the official opening of their new Cultural Centre (Figure 10.1). This beautiful cedar building of traditional architectural style was named *Waaps Wahmoodmx* ("House of *Wahmoodmx*"), to honour Chief Johnny Clifton, *Wahmoodmx*[2]. It stands as a jewel in the middle of the village, right at the front above the shoreline, symbolically representing the cultural traditions of hundreds of generations. The Gitga'at have had a long history of feasting (see Campbell 1984; Miller 1984; Anderson 1984, 2004). As in the past, guests from outside of the community were invited to attend the *Waaps Wahmoodmx* feast, to serve as witnesses, and were seated at the head table. Band members living or working away from the village often return to participate and contribute. The Gitga'at, like other North Coast peoples, have a matrilineal clan system, with three main Gitga'at Clans—Eagle, Raven, and Blackfish (Killer whale)—and descendance through the mother's line. The clan members, led by the head chief or matriarch, usually plan, organize, and host feasts. The men of that clan provide the fish and game to be served, and the women provide the preserved food, bread, cooked vegetables, and desserts, set the tables, and serve all the guests. Members of other clans may also be asked to contribute or may volunteer food or services; they are usually acknowledged for their contributions with a gift.

In the 2003 *Waaps Wahmoodmx* feast, many community members provided the food: dozens of crabs cooked in two enormous pots, a huge halibut over six feet long, cut up, and fried; kippered spring salmon; many kinds of salads; several kinds of bread, served with butter and jam; plates of cantaloupe, honeydew melon, and watermelon; cake, mints and saltwater taffy; coffee and juice. Before the feast, the children and youth of each clan performed dances and then, the House was formally opened with a prayer and a song of welcome. The guests were each announced as they entered, and were seated by designated hosts. Individual women had responsibility for each table, providing table settings—China

---

[1] This section was written by Nancy Turner, with important contributions from Belle Eaton, Colleen and Gideon Robinson, Judy Thompson (Edosdi), Cam Hill and Eva Ann Hill; all but Nancy Turner and Judy Thompson are Gitga'at community members. Belle Eaton and Colleen Robinson are Lucille Clifton's granddaughters.

[2] Sadly, less than a year later, in April 2004, Chief Johnny Clifton, the last of the old-time Tsimshian hereditary chiefs died at the age of 86 after a short illness.

plates, cups, elegant serving dishes of crystal and fine porcelain, and cutlery—from their own households. Specially designed linen napkins and place mats were given to each guest to commemorate the occasion.

Figure 10.1. Gitga'at community of Hartley Bay, British Columbia, showing the new *Waaps Wahmoodmx* Cultural Centre on the left of the photo (natural wood building)

Following the meal, the special guests were invited, each in turn, to speak and offer their appreciation and congratulations for the opening of the Cultural Centre. The guests affirmed historical connections and ties with the Gitga'at. Their formal feast names were pronounced publicly and ceremonially. The guests received bags, T-shirts, and other commemorative items, as well as apples, bananas, and oranges to take home with them. At some feasts people bring their own dishes, and take home stew and other food in their own pots afterwards. [In Haida, this practice is called *"K'aaw k'iihl"* (Turner and Wilson 2003).]

Contemporary feasts of Hartley Bay are directly connected with those of the past. A generation earlier, Chief Johnny Clifton's mother, Lucille Clifton, was renowned for the feasts that she, as Eagle clan matriarch[3], hosted with the other Eagle women for the entire village. Her main feast was around Thanksgiving time (in October), after she returned from the Gitga'at salmon camp at Old Town, and when most of the people had returned to the village after being away at the canneries or fishing during the summer months. Lucille's feasts featured largely traditional products of the land and sea from within Gitga'at territory— much of the food she prepared with her own hands.

All of the present day Gitga'at elders remember these feasts; they not only attended them as children and young adults, but also participated in harvesting and preparing the food. For example, Lucille's granddaughters Colleen Robinson and Belle Eaton and their siblings remember helping their grandmother pick blueberries (*wo'oksil)* and crabapples (*moolks*), and preserve salmon at Old Town, where they stayed with her when their parents were away working. They carried water and firewood for her, and did the other chores that needed to be

---

[3] Lucille Clifton was born in 1876 and died on 6 September 1962 at the age of 86. She was married to Heber Clifton and together they had nine children; Johnny Clifton was the youngest.

done. They watched and participated as she prepared the salmon for smoking, the salmon eggs for "caviar" (*üüskm laan*), harvested the edible inner bark of trees (*ksiiw*), and whipped the oulachen grease (see Turner et al., Chapter 1, this volume) for storing the lightly cooked *moolks* (crabapples). Now *they* are the ones preparing this food and advising others about food to be served at feasts.

As far as people recall, the menu for Lucille's fall feasts remained more or less the same, although it varied depending on availability of the different foods. Half-smoked coho (*ksü ts'aal*) was set out, and sometimes dried, soaked humpback salmon (*'wiiyuu stmoon*), along with salmon egg "caviar" (*üüskm laan*), cured and smoked, inner bark of hemlock and balsam fir (*ksiiw*), and bowls of potatoes (*sgusiit*), boiled and drained, sometimes with carrots (*'kawts, galot*) or turnips (*'yanahuu*). Toasted, flaked seaweed (*ła'ask - saxoolk'a*), and eulachon grease (*k'awtsi*) and/or seal oil (*kbaüüla*) were put out as condiments, sometimes with cooked seal meat (*üüla*). There was bread, fried or baked, with *moolks* (crabapple) jelly, or thick blueberry (*smmay*) or salalberry (*dzawes*) jam. There were strips of seal flipper, singed and cooked (*xslaxs*), and sometimes venison stew (*samimwan*). Dessert was usually crabapples (*moolks*) and highbush cranberries (*łaaya*) mixed in hand-whipped grease, as well as blueberries (*wo'oksil*), salalberries, and other berries such as *waakyil* (gray currants), mixed with sugar. Guests received oranges and apples to take home, and for beverages, they were served Salada tea and the wild Labrador tea, *k'wila'maxs*.

Some of this old-time food is no longer served today, but those who had the opportunity to develop a taste for it remember it fondly. In the earlier years, the old meeting hall could hold only about 120 to 150 people, and children were not able to come to the feast. Later, though, when a newer hall of greater capacity was built, children could come as well, and up to 200 people or more might attend altogether. In the old-time feasts, the members for each clan sat together in their own sections of the hall; nowadays people generally sit with family or friends regardless of clan. At Lucille's feast, the Ravens sat on one side, the Blackfish on the other, with the Eagles hosting and serving. The subchiefs, speakers, and those holding a special position were seated at the head table, just as they are today.

Feasts like Lucille Clifton's and the *Waaps Wahmoodmx* opening feast are critically important, not just for the food provided. Feasting enables people to enjoy and celebrate together the positive events in their lives. Of equal importance, feasts are also occasions for people to commemorate and overcome sad times and misfortunes they have endured. Community feasts promote unity, reflecting the essential need for sharing and helping each other in order for all to be able to survive. Symbolically and in reality, the Gitga'at feasts are a true reflection of such values. The diverse food that is served, much of it part of the bounty of their traditional territory that has provided for them for countless generations, is an embodiment of a rich knowledge system, commonly known as Traditional Ecological Knowledge, which is generally recognized as a knowledge-practice-belief complex relating to peoples' use of and relationship to their environment (Berkes 1999; Turner, Ignace, and Ignace 2000).

Detailed, practical local knowledge of the foods and other resources that sustain the people, and in particular, how to harvest, process, prepare and serve them, is perhaps the most obvious component of Traditional Knowledge underlying Lucille's feasts as well as, to some extent, the feasts of today. There is a great deal of traditional ecological knowledge about berries and other plant foods that is represented in the feasts. People had to know the best time to harvest the foods, and the best places to get them. For example, wild crabapples were one of the most significant types of plant foods for the Gitga'at and this is reflected in the fact that several major varieties of these are named and recognized by them. Other knowledge, including care and sustainable use of resources, is also reflected in the feasts, and, as well, this knowledge is perpetuated in the community through these feasts, among other ways. Within Gitga'at territory, some places and resource gathering sites, like crabapple and highbush cranberry patches, were owned by Chiefs and other individuals within a specific clan. It was their responsibility to look after these sites, and to oversee how they were used by and shared with the people.

Social institutions, including the family and the clans, serve as a culturally prescribed means of ensuring that everyone in the group is valued and provided for. These institutions permeate every stage and aspect of the feasts and their preparation, including the very hosting of the feasts by the clan members, the requirements for recognizing the clans and the leaders in the seating, and other protocols. The very role Lucille played in caring for and teaching her grandchildren while their parents were occupied with their work is an aspect of the institution of the family in Gitga'at life, and it carries on to this day.

Finally, the worldview or philosophy underlying the feasts is critical to their understanding and their role in helping to support and sustain the people. According to Helen and Johnny Clifton (pers. comm. 2003), giving her thanksgiving feast each year was Lucille's way of saying, "I'm glad nobody got hurt this summer. I'm glad you've looked after yourselves; I'm glad we're all back together; I'm glad about all the good food the Creator has given me." Looking upon the food as a gift of the Creator, and treating the food and all who partook of it with appreciation and respect, never wasting it, and sharing it with the entire community: these are values that Lucille herself held, and that she passed along to her children and grandchildren.

The changes in feasting that have occurred since Lucille Clifton's time are reflective of broader changes in peoples' lifestyles, in the foods consumed, and in technologies for preparing them. In the early days, for example, people picked and ate many more wild berries and wild plant foods, and consequently, these were served more at feasts. In the olden days, the elders recalled, all the women had big wooden barrels to store their crabapples and highbush cranberries in and all of them would set some by to contribute to the feasts, as well as serving them on a regular basis to their families. Clyde Ridley recalled that Lucille (his grandmother) used to put by about four or five five-gallon barrels packed with crabapples and highbush cranberries every winter. She kept some separate, and some mixed together, for feasts. Originally, people stored berries by cooking them in large pots to a jam-like consistency, then spreading them out onto skunk cabbage leaves to dry. Clyde remembered that they had square cedar-wood trays for drying berry cakes, to about an inch (2.5 cm) thick. These cakes were then rolled up, tied and put into cedar bentwood boxes, which were placed into holes in the ground lined with sand. The boxes could be dug up whenever the berries were needed. Whenever berries were wanted, people would just cut or break off a chunk and soak it in water. Alternatively, the fresh berries, especially the crabapples and highbush cranberries, would be mixed with grease and stored all winter in bentwood boxes. Nowadays, few people have the time to pick such quantities of wild berries, and when they do, they just put them in the freezer to preserve them, or they make them into preserves, jams, and jellies.

At today's feasts, the *ksiiw* and *üüskm laan* may be replaced by salads. In the place of *moolks* and *łaaya* with whipped eulachon grease, there is cantaloupe, watermelon, and other imported fruit. Handing out oranges and apples is a practice that goes back a long ways, however, probably as long as these fruits have been available in any quantity. Most importantly, the role of the feast in promoting community pride and cohesion is for the Gitga'at, as strong as ever.

THE CELEBRATIONS OF YOUTH IN NEWFOUNDLAND AND LABRADOR[4]

This section describes the way young people in Newfoundland and Labrador celebrate special occasions and how their celebrations are shaped by the cultural heritage of Newfoundland and Labrador and the social, economic, and environmental restructuring that has impacted that province in recent years. These findings are part of an ongoing study of youth in Newfoundland and Labrador that began in 2001. Focus groups with youth and adult key informant interviews were utilized to explore how changes within the community are affecting the health and well-being of youth in a community in northern Newfoundland. Eleven focus group interviews were conducted with male and female youth between the ages of thirteen and twenty-four, and nineteen one-on-one interviews were conducted with key informants who were familiar with the youth of the community from their perspectives as parents, teachers, health officials, and others who worked with the youth. The study also included a mini-ethnography which involved several visits to the community by members of the research team in order to understand contextually the narratives provided by the research participants. In addition to the above, a focus group and two one-on-one interviews were conducted to focus specifically on the youth's experience of celebrations and to add to the data already collected. The two one-on-one interviews were conducted in Labrador, and not in the community where the larger study occurred. This section uses data from the focus group and interviews that focused on celebration and food, in addition to relevant information collected from the larger study. Information derived from the larger study helped to contextualize the data related to food and celebration.

Here we discuss two types of celebrations: intentional celebrations and celebrations born of necessity. Intentional celebrations refer to those times in recent years when the youth participated in events held to commemorate a range of occasions, from the simple arrangements made for Friday night visits with friends to the much more elaborately planned events commemorating important occasions such as graduations, Christmas or birthdays. Celebration born of necessity is the term used to describe the way in which recent celebrations were often inextricably linked with the work that was necessary for the families and communities to survive.

The youth in our study described the activities, the food and the social context associated with the events they enjoyed which marked special occasions. Their intentional celebrations were important as relief from regular daily routine, for leisure, and critical social occasions that solidified allegiance to family and friends. Food was always a central part of these events and tended to be a mixture of products procured off the land and sea such as shellfish (crab, lobster, shrimp), fish (cod, caplin, trout, salmon), game and wild birds (moose, rabbit,

---

[4] The research team for this section consisted of Lois Jackson (PI), and researchers Susan Tirone, Catherine Donovan and Rob Hood. Others who assisted with the data collection and coding were: D. Hay, M. Howell, D. Martin, J. Ticknor and S. Taylor.

caribou, seal, turr, duck), wild berries, and food that is purchased from local stores. Our discussion uses quotes from the youth in our study to explain the celebrations in their lives:

> Last year, no the year before actually, we had a wedding. Now a wedding in my family is a really, really big event. So we had people from the [United] States come up . . . Everybody was invited there to stay there and there's cabins everywhere. So everyone just moved into them. We had a big turkey dinner, wine.

> Well my father just turned sixty in the year gone past and he's the oldest of, I don't know, about six or seven brothers. So we had a big party on the patio actually. And had other friends and relatives who played guitar and it pretty much involved alcohol and music. Just hanging out with friends and family.

> Q. What about food?
> Yeah, the barbeque was on the go. Probably some caplin and traditional food.

The land and sea have always been a rich and important source of food for the people of Newfoundland and Labrador. For generations European settlers and their descendents migrated seasonally between summer homes near the sea and winter homes close to areas rich in game, a practise sometimes referred to as seasonal transhumance (Hanrahan and Ewtushik 2001). For youth, participation in hunting, fishing, picking berries, and other outdoor activities related to living off the land, solidifies their place within a peer group and within their families and community. One such activity referred to as a "boil-up" is indicative of how traditional outdoor cultural practices shape the celebrations of youth today:

> I had a boil-up this summer actually. We went out in the boat and went to a beach and boiled some crabs and stuff. It was fun.

> In the winter when I'm home, most of my brothers are home, we'll probably pack up and take off out to (name of area) have a big boil-up and do some ski-doing, some snowboarding, take in the GT's and crazy carpets and probably get hurt. [GT's and crazy carpets are types of sleds or toboggans for sliding down snow covered hills.]

The youth described parties where alcohol was plentiful and food was not always consumed. Graduation parties, referred to as "grad fires," were important occasions for celebration and for attending these drinking parties:

> Q. So, any food during these barbeques, boil-ups, anything like that. Is that a big part or is it mostly drinking?
> Person 1: No, not at the grad fires. They're are all alcohol related . . .
> Person 2: Like if it's overnight maybe like, there'll be like chips, munchies
> Person 3: Unless you go with older people, there like (name). Now you go to one of his parties and he'll come out with probably three or four big trays of moose sausage about that long.

Hunting, fishing, picking berries, and the preserving and storing of local foods require a great deal of time and effort. As is evident in other cultures, celebration in Newfoundland and Labrador is not entirely a time or activity separated from everyday life. It often occurs as part of the activity necessary for procurement, preparation, and preservation of food. The youth in our study were well aware of the effort required to ensure their families had adequate amounts of food year round. They either participated directly or witnessed their parents and grandparents in such activities as hunting, fishing, picking berries, gardening, and preserving the food they procured. They described the communal nature of these tasks which provided opportunities for families to be together and to enjoy one another's company:

> Everyone hunts. Locals that lives around here year round, they all does hunting.

> Like I fish a lot in the summer time; mackerel fishing, a bit of salmon fishing, trouting, everybody hunts. Dad gets his moose licence, my grandfather gets a moose licence, Nan [grandmother] goes with him.

> We got a big garden, like a big potato garden and it's like, well I said, Dad got a big family. Like there's fifteen. So it's all the brothers in (name of community). It's all one big garden. So they all plants it together, all the brothers and the families and the kids and wives, plants the garden and harvests the garden together.

One young person interviewed in Labrador recalled members of his extended family who gathered for a funeral. Many members of the family lived away, and the family had few opportunities to be together. When they gathered to commemorate the passing of a loved one, they used the occasion to go "up the bay" to fish as a family:

> Mom and them has outings every now and then up the bay.

> Q. Outings? What do they do there?
> Go up trouting, or go up ice fishing.

> Q. Who does she go with?
> Well if mom finds out any of dad's family is going up, then they usually goes up with her. Like when Pop and Nan [grandfather and grandmother] died they had the whole family come home and that was in the winter . . . I wasn't home but they all went up the bay. I wouldn't go up anyway, but they all went up in the bay and had a big old time up there. A big party.

Although this family had gathered to grieve for their loved one, the occasion became a celebration of family togetherness. Gathering at their fishing cabin solidified their connection to one another and to the land. Food obtained from the sea and from the land and the activities required to obtain that food are of primary importance in the lives of the youth we interviewed. Food and procurement activities appear to provide the youth with a strong sense of their cultural heritage, and their roles as important contributing members of their family and community. However, changes in the environment, economy, and in their social relationships have in some ways affected how they celebrate, as well as the traditional food that has for many generations been central to these celebrations.

The most drastic change they described had to do with outmigration—the many family members and friends who have moved away:

Well ours (family celebrations) have definitely decreased over the last ten years I guess. I have a lot of uncles and cousins. They grew up and just went away, no work. They either go to school, go to college, or university, go away and work, sometimes you make it home, usually just holidays.

The intentional celebrations associated with family gatherings remain important for the youth, and may have increased in importance because, due to outmigration, they occur less frequently.

The ways in which youth celebrated Friday nights, and other such simple occasions have changed as well, as one person explained:

Person 1: There's more house parties now, more people's places to go. I mean, I'd prefer to go to my house than everyone go outside.

Person 2: Like when we were in high school there was no house parties because there was so many people coming, there's always

Person 3 (interrupting): Something broken

Person 4: Big fights and stuff like that, more people.

Parties with friends have become smaller as friends moved away, but they may have been somewhat safer than those held in the past on beaches and in the woods. When parties occurred out of doors people often stayed out overnight, sleeping in cars or vans because they often drank alcohol and stayed at the party site, rather than driving home:

Person 1: You're drunk and you're not going to drive home so . . .

Person 2: We just continues with the drunkenness and hitch hike home. We used to go hard.

Person 1: Well I put a mattress in the back of my van, so I just used to go in my van.

The consumption of alcohol was mentioned frequently by the youth as something they did with friends and with family. However, the youth were well aware of the need to ensure they and their friends were safe during parties where alcohol was consumed, and some key informants explained that the youth were more cautious than their parents' generation had been when they were teenagers.

Although local food and drink is an important part of their diet they also had access to food that was not locally produced. With improved road access to northern Newfoundland and Labrador, food is trucked into the area giving people more options and choices. We frequently heard that the youth ate less home grown and prepared food than in the past, even though the food available at the stores was expensive:

I don't fish. I don't agree with the fact of, if I don't eat it then, like, I'm not going to catch it. I don't see the point. Dad can get his own fish.

It tastes better, 'cause you got the road now and a lot of fresh comes in; a lot of fresh meat. We have good meat. Like years ago the hamburger meat was black here, like you'd eat it because we didn't know the difference.

Q. Did [your mother] bake more when you were younger?
Mom did but I often heard her say how everyone used to bake and now she got friends who never bake. Like they always buy their bread.

The food preferences of the youth we met were influenced by radio and TV advertisements for large, multi-national, fast food companies. Unlike youth in large urban Canadian centres, the youth in our study had limited choices of fast food and other restaurants. So, when they had opportunities to travel to larger centres visits to Wendy's and McDonald's were a very popular part of the ritual of "going to town." However, these meals were not always altogether satisfactory:

I went up the McDonald's. We hauled up to the drive through and the waiter goes "What do you want?" I said, "I wants a Big Mac, fries and gravy." They don't got no gravy here. Well lard, dyins, ["Lord dyings" is a colloquialism, expressing surprise and exasperation] McDonald's, no gravy. You'd think they'd have gravy there though hey? That's pretty stupid.

Both intentional celebrations and those that occurred along side the necessary tasks required to keep families together have changed in recent years. Most noticeable is that family celebrations occurred less frequently because of outmigration. The celebrations were an important aspect of life for the youth in Newfoundland and Labrador, not only as leisure, but also because these events solidified their connection to family, friends, and community. As well, both types of celebration provided many opportunities for the youth to engage in outdoor activities, further strengthening their connections to the land and their love of the physical environment in which they were raised. As participants in family celebrations and in activities necessary for sustaining homes and family life, the youth learned how to procure and prepare food—skills they will be able to use when they host and participate in celebrations with friends. As well, the celebrations they experienced with both family and friends helped create order and structure in their lives and provided them with opportunities to experiment with adult roles and to assume responsibility within their family and community.

FOOD AND IDENTITY[5]

Identity has been defined as the construction of a sense of self as well as the "unfolding bridge linking individual and society, childhood and adulthood" (Josselson 1994, 12). Individuals build these mental self-images based on their everyday interactions with their cultural, structural, and social environments. Identities are also shaped by personal narratives and shared metanarratives of a community (Halter 2000). Warde (1997) uses the terms communification or social re-embedding to refer to the need to connect with people who

---

[5] This section was authored by Blythe Shepard and Anne Marshall. Blythe Shepard is a research associate with Anne Marshall who headed the research team. Research involved examining the impact of social, economic, and environmental restructuring on youth living in five coastal communities in British Columbia.

share common narratives based on geographic space, traditions, aspirations, or standards. Food and the sharing of food is an important aspect of the day-to-day events shaping people's lives and thus contributes to the development of identity in young people.

A case study of a secondary school on the west coast of British Columbia illustrates the importance of food events in the construction of identity. The focal community in this study is a town of 5,000 that was once a thriving mining and fishing community. At its peak production period the mine employed over 900 people and spent $1.2 billion in British Columbia on supplies and services during the years of operation (Britton 1998). The mine officially closed in 1995 when it became no longer profitable. At the same time there was a general decline in commercial fishing. Since 1984, the commercial fishery in BC has declined (Hallin 2000). In the mid-1980s commercial fishing accounted for 1% of the economy in terms of the number of jobs and its contribution to gross domestic product (GDP). By 1999, that rate had fallen to less than half a percent (Hallin 2000). During the 1990s unemployment rates in the commercial fishery remained close to 20% or about twice the provincial average (Hallin 2000). The greatest decline in salmon stocks in this century occurred in 1998 and 1999 and, as of 1999, 16,000 fewer fishermen were employed in the commercial salmon industry than in the mid-1990s (Walters and Korman 1999). When families moved away to seek work elsewhere, the student population at the secondary school gradually decreased from 453 in 2001–02 to 368 in 2005–06 (BC Ministry of Education 2005). The loss of students has impacted school-based social and cultural celebrations. However, even with declining enrolments the school has continued to focus on youth involvement and has maintained close connections to the community it serves.

The school's community-based curriculum includes food events that provide a medium for celebrating youth's accomplishments, for acquainting young people with and honouring their cultural background and heritage, and for strengthening youth's sense of self as a community member. Homemade potluck dinners surround theatre productions and sports events. Multicultural celebrations provide families with opportunities to share food. Aboriginal students discover how to cook local plant and animal foods through their Elders. Monthly suppers and breakfasts held in the school auditorium bring senior citizens, families, and youth together. Such celebrations assist youth in their search for some natural rootedness or belonging associated with sharing with people around them and strengthen relationships among families and communities.

In fact, food plays a critical role in celebrating youth's accomplishments and in bonding youth with the school environment. "[T]here was always food, always potluck food that parents and kids had made arrangements to bring in that surrounded our school celebrations" (educator). Students enjoy their role as guests of honour.

> We would do this [theatre] production, we would be so excited . . . and then we'd have this wonderful food right after . . . in the gym . . . with our family, everyone's parents . . . It felt great, I think because we felt like we belonged, like we had something to offer the community . . . and the community valued our work (student).

The First Nations students who took part noted how celebrating their accomplishments reinforced their commitment to stay in school and to be involved.

For me and for some of my friends, we're more likely to stay in school, more likely to finish school if we feel we're wanted. This school does that . . . They do things like organizing potlucks to recognize our achievements . . . that's what we need (First Nations student).

Through providing students with positive experiences in a caring environment, youth develop pride in their identity. Schools in small communities form the hub of a wheel, with spokes to family and community services. When lines of influence run in both directions, schools attract family and community agencies into learning and social developmental partnerships mobilizing the capacity building of the public. Positive youth-context relations ensure that young people have the nurturing and support needed for healthy development (Lerner et al., 2002).

The transformation of schools into places of human attachment requires routines and social activities that bring youth and adults together to develop ritual practices and self-understanding as active, responsible members. When students at this school wanted to develop social activities to build solidarity among the diverse student population, they asked their principal for assistance in staging a multicultural day.

It was a one-day celebration. There were workshops all through the Friday and in the afternoon the gym was set up with the displays, both active and passive. Some of the active displays, for example, were fashion shows put on by people from the different cultures or countries, karate, dancing, and that sort of thing, and then in the evening there would be another demonstration. There was also food—ethnic food—dishes that you couldn't find anywhere else in the area (educator).

When resource industries flourished, residents from many countries created a diverse community. Eight years ago, twenty-three different nationalities were represented at the school's first multicultural day and about 700 people attended. One youth remembers the joy she experienced in cultural sharing.

It was fun. There was noise and colour and the smells! It reminded me of an outdoor market in some far off country . . . Really, you hardly knew that you were in a fishing village . . . I started to make friends outside my little circle—to learn more about others. When I think back to that time, those opportunities made me proud of my ethnic identity . . . but it also made me appreciative of other ethnicities (female youth).

The educators at the school collaborated with students and parents to promote social functions in which multiple perspectives were heard, reflecting ethno-cultural, linguistic, and racial diversity. The experience serves as an example of Lerner *et al.*'s (2002) assertion that when youth develop cultural awareness they also develop caring values that can reduce inequities in their surroundings.

Aboriginal youth often experience considerable conflict between the expectations of their cultural group and the norms of the dominant society (O'Neil 1986). School food programs run by First Nations teachers provide one way for aboriginal youth to stay connected to their culture. Elders instill pride of self and of group by sharing their knowledge of food gathering and preparation with young people.

[S]everal people who knew the traditional way of barbequing salmon, were there to show the children . . . how to split the salmon, how to set up with sticks that are stuck into the ground over the fire. While it was cooking, Wata, who is a medicine woman, showed them the flora of the island that they were on, nutritional and medicinal qualities, and they did some picking of some of those foods, and then they had a feast (educator).

Food serves as a powerful symbol of cultural values and identity. Being First Nations, like my grandmother and aunties, they teach me that these are salal leaves and they are so knowledgeable about living close to nature. Learning about my culture . . . about traditional ways of doing things, I feel . . . there is a connection for me to earlier times and what it means to be First Nations . . . It helps me to see who I really am (First Nations youth).

Sharing food has always played a very functional and complex role within First Nations social organization. Aboriginal students, who make up 44% of the student body at this school (BC Ministry of Education 2006), have underscored the importance of food culture in that environment. As the First Nations counsellor stated, "You ingest information best when you ingest it with food."

Food events provide an opportunity for students to take on leadership roles by cooking and serving school lunches once a week.

It was sort of the one time in the week that everybody was together around food. We had several programs that cooked on a regular basis, would do luncheons for invited students and staff. We had obviously a foods and nutrition program that went from nine to twelve and they did that sort of thing quite a lot. Young parents program would put on luncheons, and there were a number of FN programs going on in the school (educator).

School-family partnerships built on social interaction can promote positive healthy development in youth. Although economic and human capital may have been stretched in the community, educators have relied on the abundance of social capital (Coleman 1988) inherent in already existing community relationships. Teachers and administrators have cultivated a strong sense of place by providing opportunities and spaces for parent and youth involvement. Families and children occasionally share dinners at school.

We had spaghetti suppers; we had at least one of those every year. It wasn't a fundraiser, just a time for families to come together. Parents and kids would set up tables in the gym, and people would come in and pay by donation and have a spaghetti supper (School counsellor).

School related activities that provide opportunities for inclusion in the larger community may reduce the growing alienation of youth. "Knowledge of place is . . . closely linked to knowledge of the self, to grasping one's position in the larger scheme of things, including one's own community, and securing a confident sense of who one is as a person" (Basso, as cited in Osborne 2001, 43). School leadership roles strengthen intrinsic work values and foster self-exploration with respect to values and one's role in the community (Lerner et al. 2002). Furthermore, developing relationships with a variety of adults and peers can promote prosocial norms and enhance the amount of social capital in the community.

Food events involve social sharing which carries symbolic and cultural significance. Because of its symbolic importance in childhood and its central role in every culture, food events serve to embody and organize our social relationships. Food connects youth to community because rituals around food provide youth with a basis in a world that is increasingly materialistic and de-ritualized (Dorsa 1995). Food celebrations also seem to increase openness to multiculturalism and respect for nature and ecology. The experiences of youth and adults in this coastal community underscore the central importance of their identity-building opportunities.

CONCLUSION

Our three accounts of celebration and food provide evidence of how participation in family and community activities is critically important if people are to find a sense of meaning in the places where they live. All of the people we worked with or interviewed and the communities in which they live, have been affected by change due to major restructuring. The circumstances faced by our study participants are not unique; other Canadians and people throughout the world encounter change as well. Opportunities to celebrate with food provide people everywhere with the means for connecting on many levels. They help to solidify ties within families and communities, and to recognize and honour those we value. As change impacts the food available for celebrations, the occasions that people value, and the context in which they occur, further studies are required to explore how celebrations may change in changing times, and if important opportunities for connecting are being created, or lost. Celebrating with food connects people with their past and fosters an appreciation for the gathering and dissemination of knowledge. Knowledge of the past and one's ability to connect with family and community are critically important for building resilience in times of change.

ACKNOWLEDGEMENTS

Many Gitga'at community members shared with us their memories of Lucille Clifton and her fall feasts and stories. We are especially grateful to the late Chief Johnny Clifton and to Helen Clifton, Colleen and Gideon Robinson, Belle Eaton, Archie Dundas, Elizabeth Dundas, Reverend Ernie Hill Sr. and Marjorie Hill, Ernie Hill Jr., Lynne Hill, Cam Hill and Eva Ann Hill, Clyde Ridley, Alan Robinson, Tina Robinson, and Dick and Mildred Wilson. We also thank Pat Sterritt, former Chief Councillor and the Gitga'at Nation Chief and Council for permission for *Coasts Under Stress* researchers to participate in research with the Gitga'at Nation and to attend the *Waaps Wahmoodmx* feast in June 2003.

We thank the people of northern Newfoundland, who gave so generously of their time to participate in the study, and those who transcribed the Newfoundland tapes: P. Harris, J. Sheppard-Wells, C. Matthews, and S. Williams, S. Wheaton, and C. Jones.

As well, we thank the educators, students, and community members who took the time to share their experiences of food events within the local secondary school of Vancouver Island.

Thank you also to Dr. John Lutz, and to *Coasts Under Stress* MCRI research project (Rosemary Ommer, PI), and SSHRC research grant (# 410-94-1555) for supporting our research.

In: Resetting the Kitchen Table
Editors: C. C. Parrish et al., pp. 161-175

ISBN 1-60021-236-0
© 2008 Nova Science Publishers, Inc.

Chapter 11

CHANGING PATTERNS OF HOUSEHOLD FOOD CONSUMPTION IN RURAL COMMUNITIES OF ATLANTIC CANADA

Shirley M. Solberg[1], Patricia Canning[2] and Sharon Buehler[3]

[1]Nursing and Women's Studies, Memorial University of Newfoundland, Canada
[2]Education, Memorial University of Newfoundland, Canada
[3]Epidemiology in Community Health and Humanities,
Memorial University of Newfoundland, Canada

ABSTRACT

It is not unreasonable to expect that many of the foods served on the tables of rural households in coastal communities differ from those of a few decade ago given the changes in technology, transportation, environments, and social organization that have occurred in coastal areas of North America. Food consumption patterns at the household level are sensitive to these changes as well as to other structural and individual factors such as globalization and availability of particular foods, food product advertising, income, health promotion initiatives, and individual and cultural food preferences. Rural communities undergoing economic and social restructuring may be especially vulnerable to food security issues as these communities have reduced goods and services and experience higher levels of unemployment. In this chapter we explore how food consumption patterns in selected rural communities in Newfoundland and Labrador on the east coast of Canada have changed. We draw on data from surveys conducted in 1,470 or 70% of the households in three rural areas on the Great Northern Peninsula of the island that vary in size and economic diversity, and examine changes in food consumption patterns over a thirty-year period in one of these geographic areas. The latter helps to establish just how much household patterns of food consumption have changed and in a few instances how little has changed. More households are now using vegetable oil for cooking and have less use of lard and salt pork fat. The consumption of traditional foods such as salted beef and fish have also decreased. Fruit consumption of apples and oranges has not increased appreciably. Not many people grow vegetables or keep animals for household food consumption. We also examine some of the factors that may explain these changes and some of the constraints that households in rural communities experience in achieving food security. The chapter helps to provide an understanding of the complexity of food security and some of the challenges facing rural communities.

INTRODUCTION

Patterns of household food consumption, i.e., what people eat, at the household level within a community, can inform us about identity, social relationships, culture, politics, health, resilience, and food security in these communities (Chamberlain 2004; Wallace 1998). Yet patterns of food consumption are very complex and often poorly understood (Booth 1998). The type of food consumed and the place of consumption, i.e., at home and location in the home, or outside the home, define these patterns (Dickinson and Leader 1998). Food consumption patterns are dynamic and continue to change in response to a wide array of factors (Canada, Health and Welfare 1990; Regmi 2001). Structural factors such as globalization in relation to food distribution, technological developments in food production and preparation, health promotion and nutrition, and the knowledge of nutrition in chronic diseases, beliefs about food and risks associated with particular foods, and food availability in specific areas all influence these patterns (Schluter and Lee 1999). Kealey (Chapter 12, this volume) illustrates how the state worked through nutritional policy and health professionals to influence people to change dietary practices.

Individual factors such as taste and food beliefs also have an impact on what people eat. Particular foods are associated with gender and identity (Counihan and Kaplan 1998; Scholliers 2001). Culture influences local diets in that some food is more closely associated with regions and population in the world than others. This cultural construction around food is used to define "traditional food" (Amilien 2001). Even within geographic regions there may be local variation. For example in some areas in Newfoundland and Labrador where this study took place, squid is considered a delicacy while in another area it is seen as bait to get more desirable fish on the dinner plate. Economic resources within the household also influence food consumption patterns and taste and may be the most important influence on food consumption within a household (McIntyre 2003; Ralph 1998). The type of food consumed is only partly individual choice, because it is a choice constrained by a wide array of factors. These factors are determined at different levels that Nathoo and Shoveller (2003) defined as micro- (individual), meso- (community), and macro- (socio-cultural and policy forces) levels.

Changing food consumption patterns at the household level can also provide information on some of the transitions that are taking place in communities (Schmidhuber 2003). These transitions may indicate changes in supply and distribution of food as well as how people have modified their diets in response to a host of factors. They also inform about new services offered in communities, such as restaurants and fast food outlets. The coastal communities that we have studied within this project have undergone a great deal of change through the restructuring processes that have taken place. These restructuring processes have affected the social systems, (e.g., education and health), economic systems and social security, and the environment. Other factors affected have been communication, transportation, and the social organization of the communities. All of these have ripple effects on food consumption patterns in the household and the community. Within the wider project we attempted to document some of the restructuring that has taken place and to examine the link to human and environmental health. The various forms of restructuring we have studied have had an impact on human health, that is, the health of individuals, families, and communities.

In examining changes in human health we took a health determinants approach and attempted to look at these various determinants. One of the determinants included most relevant to the focus of this book is food. In keeping with the World Health Organization's *Social Determinants of Health* document (Wilkinson and Marmot 2003) we understood food to be affected by cultural, social, and environmental changes that are taking place in our communities. Our approach to this topic, and what is represented in this chapter, was to look at how current food usage and availability was being impacted. We did this by focusing on changing household food consumption patterns. We performed a large-scale survey of households to examine this change. Part of our research builds on the work of a health survey done in three communities within our study area almost thirty years ago (Marshall 1975). This approach has allowed us to connect the past with the present and give a longer view of change, an approach that has been important to our entire research project.

The purpose of this chapter is to illustrate, through the use of a case study into nutritional change within selected coastal communities in Newfoundland and Labrador, how food consumption patterns are changing within this context. We explore some of the factors that may account for these changes on the east coast of Canada, and provide a greater understanding of food security issues in light of these findings. The study allows us to examine how factors like health, culture, and resilience have an impact on these changes. It is a detailed look at just how the kitchen table is being reset and why. In the chapter we will present a comparison of food consumption patterns in six communities, grouped into three geographical areas, in the province. Within one of the areas we will compare changes that have occurred over a thirty-year period. The particular foods included in the study were milk, fats, meat, seafood, homegrown vegetables, and fruits. In the category of meat we focused on game, but we also considered the fish and shellfish that was readily available and traditionally consumed within the province. Likewise, under fruit we asked about locally occurring berries gathered by the people in the communities. Looking at locally available food shows how the informal economy contributes to food security and this approach complements the work of Ommer, Turner, MacDonald, and Sinclair (Chapter 8, this volume).

THE STUDY COMMUNITIES

All the communities in the study are located on the Great Northern Peninsula of Newfoundland and Labrador. Area 1 consisted of the communities of St. Paul's, Cow Head, and Parsons Pond on the western side of the coast in fairly close proximity to Gros Morne National Park. These communities depended on the cod fishery and were greatly affected in 1993 when the cod moratorium that had been put in place in other parts of the province in 1992, was extended to the West Coast. The area has had high rates of unemployment and has been characterized as geographically and economically marginal (Felt and Sinclair 1995). Much of the work is seasonal and in the service sector. The close proximity to a national park means that some of the seasonal industries are aimed at the tourist trade in the form of providing food and accommodation or entertainment. While a number of families in recent years have moved to other parts of Canada where chances of employment are improved, a common pattern is for the male head of the household to go away for varying periods of time for employment, while the female spouse maintains the household in the home community.

Area 2 lies further north on the peninsula and consisted of the communities of Port aux Choix and Port Saunders. These communities were home to larger offshore fishing fleets and while able to diversify from the cod fishery to the crab and shrimp industry, have had their share of economic difficulties in more recent years. Finally Area 3 consisted of the community of St. Anthony. While fishing is important to this community, it is also a service centre for other areas of the Great Northern Peninsula and has a more mixed economy than the other two regions. It is located on the tip of the peninsula. These three areas are therefore diverse as to the timing and extent of restructuring, particularly within the fishery that has taken place. Table 11.1 contains a summary of selected demographic and economic characteristics of these communities.

In the fall of 2004 we carried out a health survey in households in these three areas and included a food consumption questionnaire where we asked about household frequency usage of common foods by asking if they were consumed in the household "often," "now and then," or "never" along the three time dimensions of the present, five years ago, and ten years ago. This approach was in keeping with data collected in 1974 in Area 1 when the same format was used. The same food items were included as well. The only differences in data collected between the two time periods was the addition of "fast foods" and some questions on food availability and shopping in the 2004 survey. In 1974 when the original food consumption patterns survey was done the combined population of these three communities was 1,545 in 306 households (Marshall 1975). In the latest Statistics Canada survey collected in 2001 the number of households had increased to 567 while the population has decreased to 1,267. There are more single-family households than had been the pattern previously.

The response rate for the 1974 survey was 274 or 81% of the eligible households in Area 1 and for 2004 it was 1,470 or 70% of the eligible households in all three areas. The availability of the nutritional data set from 1974 allowed us a unique opportunity to build on this work in these same communities. Older people in the communities remembered being respondents in the 1974 surveys and commented on the study to the research assistant.

Table 11.1. Demographic and Economic Profile of the Study Areas

Indicator	Area 1 3 small communities	Area 2 2 medium communities	Area 3 1 large community
Population 2001	1,268	1,822	2,730
Number households 2001*	475	650	970
Population change from 1996 to 2001	-18.9%	-9.6%	-8.9%
Employment insurance incidence 1998**	68.6%	63.2%	26.9%
Average husband-wife income 1999**	$36,000	+$47,850	$53,600
Social assistance incidence 1998**	31.9%	9.3%	9.8%

* Canada, Statistics 2004
** Newfoundland and Labrador Community Accounts, n.d., well-being

CHANGES IN FOOD CONSUMPTION

The findings from the 2004 survey show that there are some similarities and some differences among the three areas in terms of the types and frequency of food consumed. This section highlights some of those comparisons. It is divided into the different food groups examined and contains key findings from both the older and more recent surveys with a comparison of what has been happening in Area 1 over the timeframe covered by the two surveys.

Milk. Three types of milk products were examined at both time periods; homogenized milk, canned evaporated or condensed milk, and powered milk, or as the first two were referred to in the vernacular "fresh" and "tinned" milk. The latter was used in tea and coffee and even today if you have a cup of tea or coffee in the area and request milk you might be asked, "Do you want fresh or tinned?" Evaporated condensed milk was also used as an infant formula and an important function of the public health nurse was to ensure the mother knew the correct formula for this type of milk based on the age of the infant. Although not recommended as an infant formula, evaporated condensed milk is still given to very young infants. Tinned milk was a staple because it has the advantage of surviving long periods of transportation, as unopened it does not require refrigeration and even opened can resist spoilage if kept in a cool place, unlike homogenized milk that requires refrigeration for transportation and storage. Although powdered milk was promoted as a viable alternative to homogenized milk in nutritional circles it never had widespread consumption.

Over the past decade there has been little change in the "often" category of the use of fresh milk in Area 1, however usage has increased in Area 2 over this same time period and even increased more in Area 3. The use of canned evaporated milk has remained fairly stable in the three areas, but always had a higher consumption in the households in Area 2. Powdered milk now as in the recent past had a low frequency of usage and that has not changed by area or across time. Table 11.2 illustrates the milk consumption patterns for the three areas in the recent past.

Table 11.2 Percentage Change in Milk Consumption Patterns by Area, Frequency and Time

Type of Milk/Frequency	Area 1			Area 2			Area 3		
	Now	5 yrs.	10 yrs	Now	5 yrs.	10 yrs	Now	5 yrs.	10 yrs
Fresh milk									
Often	68	65	65	71	68	66	84	73	66
Now and then	21	24	22	21	24	25	10	18	21
Never	11	11	13	8	8	9	6	9	13
Evaporated milk									
Often	50	51	52	68	69	70	56	59	60
Now and then	27	28	27	22	21	21	27	26	25
Never	23	21	21	11	10	9	17	15	15
Powdered milk									
Often	4	4	3	4	3	5	3	4	5
Now and then	8	11	12	12	16	16	11	14	15
Never	88	85	85	84	81	79	86	82	80

Looking at Area 1 where comparative data are available illustrates that patterns of milk consumption have changed markedly from the 1974 Health Survey to the present day. In the former period "tinned" milk was definitely the milk of choice in households with 91% of households reported using it "often" and hardy anyone said they never used it. Thirty years later consumption of evaporated milk had declined, but 50% of those surveyed report frequent consumption of this product. "Fresh" milk or homogenized milk consumption was used "often" by just over one-third of the households at that earlier period. Use of homogenized milk has greatly increased so that now, up to 70% of households report frequent usage. No doubt refrigeration and transportation have been contributing factors to the more frequent use of this milk product. Despite being promoted by health professionals, the household use of powdered milk just never caught on.

Fats. Under this category we tried to establish the use of lard and salt pork fat versus vegetable oil for cooking and consumption of butter versus that of margarine. Nutritionists have moved back and forth on advice whether or not to eat butter or margarine preferentially (Wallace 1998). In the recent survey the use of lard for cooking and baking is low in frequency of usage and has shown a slight decrease across the research areas. The finding that does differ from the earlier survey to the later one is in the category of "now and then," with Areas 1 and 3 reporting that around a third of households report this frequency of consumption of lard, while in Area 2 it has a low occasional use. Vegetable oil is the fat of choice for cooking and as would be expected given the findings for lard, households in Area 2 use it with more frequency on an "often" basis. Around half of the households in Areas 1 and 3 use vegetable oil frequently. Salt pork fat has traditionally been used as a fat for cooking in the province, but shows a decrease across the board in frequency of usage. Margarine has high frequent usage in all areas and has increased somewhat in frequency in Area 2 while remaining stable in the other two areas. In contrast butter is used less frequently by all households and has a lower frequency of usage in Area 2 (see Table 11.3 for changing fat patterns). Various forms of margarine have a longer shelf life than butter. In addition it is not as costly as butter to purchase.

Fat consumption of the foods measured has changed in the thirty-year period in Area 1. The frequent use of salt pork for cooking has decreased by 64%. Lard has shown a more marked decrease from 71% frequent usage by households to 5% at present. Vegetable oil is the fat of choice for cooking and 54% said they use it often as compared with the 16% of households indicating this usage thirty years ago. Butter had low frequent usage in 1974 at 6% compared with the 89% of respondents who said they used margarine at that time.

Meat. Salting beef in brine was a way of preserving the meat so that it could keep over a longer period. This product is simply referred to as "salt meat." A traditional Newfoundland dinner is a "boiled dinner" with salt beef and vegetables, mainly potatoes, carrots, turnips, and cabbage, and with peas pudding and a flour pudding called a "duff" cooked in the same pot. It is similar to a New England boiled dinner or the "bouille" in Quebec, although some regional differences are noted between these and the Newfoundland version. There are some variations in the meal even within the province. A salt beef dinner was frequently made for Sunday dinner at noon. However, much nutritional education has been conducted around reducing salt in the diet as one measure to combat the high rate of cardiovascular disease in the province, in particular the high rate of hypertension (Newfoundland and Labrador Centre for Health Information 2004). Salt beef is more frequently consumed in Area 2, followed by Area 1, and less so in Area 3, however consumption in all three areas have remained stable over the decade.

**Table 11.3. Percentage Change in Fat Consumption Patterns by Area,
Frequency and Time**

Type of Fat/Frequency	Area 1			Area 2			Area 3		
	Now	5 yrs.	10 yrs	Now	5 yrs.	10 yrs	Now	5 yrs.	10 yrs
Lard									
Often	5	6	7	3	5	7	2	3	5
Now and then	33	37	39	16	14	15	24	29	30
Never	62	57	54	87	81	78	73	68	65
Vegetable oil									
Often	54	52	50	74	70	69	46	43	42
Now and then	43	44	45	23	24	19	47	48	47
Never	3	4	5	3	5	12	7	9	12
Butter									
Often	6	6	7	5	5	8	6	6	7
Now and then	19	22	22	11	14	14	17	16	17
Never	75	72	71	84	81	78	77	78	76
Margarine									
Often	89	90	89	90	89	85	80	80	80
Now and then	9	8	9	7	8	11	17	16	17
Never	2	2	2	3	3	4	3	4	3
Salt pork									
Often	13	14	17	13	17	23	7	12	18
Now and then	57	59	57	37	40	36	42	46	46
Never	30	27	26	50	43	41	51	42	36

Apart from salt beef we looked at the consumption of "fresh" beef (i.e., roasts or steaks). Beef is consumed more often on a frequent basis in Areas 2 and 3, with a marked difference in Area 1 and this pattern has held over the period of study. We also included wild game. The greatest source of meat comes from game in the form of moose. Hunting, particularly big game hunting, is an important activity on the Great Northern Peninsula (Omohundro 1999). While the residents enjoy the recreational aspect of the "hunt" it also provides an important source of meat. People in households in Areas 1 and 2 rely heavily on this game for meat. In fact over 70% of people in these communities consume it often. This is not so in Area 3 where only one-third of the households report frequent consumption; a pattern that has not changed appreciably over the decade studied.

One of the means of preserving moose is to bottle it and around 88% of the households use bottled or jarred moose meat at present, 68% "often," and 42% "now and then." A research assistant living in the area during the period of data collection in the spring of 2004 remarked to a participant on the big difference between beef and moose consumption and the response was, "Sure, you can do anything with moose that you can with beef—any recipe." Residents would talk about the nutritional advantages of moose over beef, in particular that it was low in fat. They also spoke of it as being more flavourful than the beef that they were able to obtain in stores and supermarkets.

Raising animals for home food consumption is one way to supplement the household diet. Hardly any of the households or communities raises chicken for consumption or to produce eggs. Frozen chicken is now readily available on the island and at a more reasonable cost than other types of meat. People also hunt sea birds as a form of meat (See Montevecchi et al. Chapter 7 this volume). Table 11.4 highlights some of meat consumption over the past decade.

Table 11.4. Percentage Change in Meat Consumption Patterns by Area, Frequency and Time

Type of Meat/Frequency	Area 1			Area 2			Area 3		
	Now	5 yrs.	10 yrs	Now	5 yrs.	10 yrs	Now	5 yrs.	10 yrs
Beef									
Often	17	17	17	48	45	43	43	44	44
Now and then	60	60	61	43	45	45	43	44	44
Never	23	23	23	9	10	11	4	4	4
Salt beef									
Often	35	37	38	44	46	48	16	23	29
Now and then	56	55	55	40	40	39	63	60	55
Never	9	9	7	16	14	13	21	17	16
Moose									
Often	72	71	71	77	77	76	34	34	35
Now and then	26	26	25	17	17	18	49	49	47
Never	2	3	4	6	6	6	17	17	18

Comparing the 2004 survey data with the 1974 data shows that consumption of fresh or frozen beef has decreased in frequency in that in the earlier survey 48% of households consumed it often. Salt beef consumption too has decreased somewhat from 74% responding "often" in 1974 compared with 35% at present. In contrast moose consumption has increased over the thirty-year period; up from 44% who consumed this game often in 1974. More people also preserve moose by bottling it now than did thirty years ago. There was also a change in the number of households who raise chickens for home consumption. Keeping chickens for food and egg production is almost non-existent in Area 1 whereas three decades ago 10% of households indicated they raised chickens for food and 23% produced farm eggs for household consumption. Even at that time a 14% decline in raising chickens was noted from ten years earlier, i.e., early 1960s. Omohundro (1994) in his fieldwork on the Great Northern Peninsula in the 1980s noted that in general animal husbandry of any kind was diminishing as a household production method.

Fish and Shellfish. A number of our questions involved the inclusion of various types of fish or shellfish in the diet: fish such as cod (including salt cod), herring (including salt herring), salmon (wild and farmed), seal, and shellfish (lobster and crab). Shellfish are an important commercial species in the area. The most frequently consumed fish in all three of the study areas is fresh or frozen codfish, followed by salt cod, salmon, and then herring with fresh and salted being fairly even. Area 2 stands out as having a higher consumption of these types of fish than the other two areas. However there is a slight decrease in consumption of all types of fish over the past decade. Lobster, crab, and seal have a much lower frequency of consumption rated as "often," even when we out in the condition "in season." They are all consumed "now and then" at a high frequency with Area 1 having the highest consumption in this category. Table 11.5 summarizes changes in seafood consumption.

Table 11.5. Percentage Change in Seafood Consumption Patterns by Area, Frequency and Time

Type of Seafood/ Frequency	Area 1			Area 2			Area 3		
	Now	5 yrs.	10 yrs	Now	5 yrs.	10 yrs	Now	5 yrs.	10 yrs
Cod									
Often	26	28	29	41	55	64	25	33	49
Now and then	70	69	67	51	41	33	68	62	47
Never	2	3	4	8	4	4	7	5	4
Salt cod									
Often	19	20	20	31	40	48	12	19	30
Now and then	74	73	73	46	46	39	66	64	54
Never	7	7	7	23	14	13	22	17	16
Lobster									
Often	12	13	13	22	23	25	3	4	8
Now and then	81	81	81	65	65	62	48	49	46
Never	7	6	7	13	12	13	49	47	46
Crab									
Often	9	9	9	18	18	21	10	9	13
Now and then	77	77	76	56	56	54	63	62	57
Never	14	14	15	27	27	25	27	29	30

With respect to the consumption of seafood between the first and second surveys, 28% of households said they ate lobster "often" in season thirty years ago compared with at present (12%), but crab consumption has increased in that 9% of households said they ate it frequently at present as compared with 1% of the households that indicated frequent consumption at the earlier survey. In examining the consumption of fish, i.e., herring (fresh or salted), salmon, and cod (fresh or salted), included in the two surveys the frequency of having any of these on the dinner table has decreased with the consumption of salted fish showing the greatest decrease.

Garden Vegetables. Research by Omohundro (1999) in the 1980s was in part influenced by his observation of gardening patterns on the Great Northern Peninsula. He noted that vegetables grown in gardens had contributed to the nutrition of the people living in the area particularly prior to the 1950, i.e., pre-confederation with Canada. In our recent survey in 2004 we asked our respondents about the use of their own garden produce for household consumption. Around 30% said they consumed their own homegrown potatoes "often," a decrease in about 10% from the previous ten years. Likewise there was a decrease in consumption of other household garden produced vegetables, such as turnips and cabbage. Potatoes are the most frequently grown and consumed crop in all three areas, their production and consumption has decreased over the past decade, but consumption varies by area. Area 3 has the highest level of home garden produced potato consumption and Area 2 the lowest. The production and consumption of root crops such as turnips and beets, and cabbage are fairly similar and lower than that of potatoes. Garden food production as a contribution to the informal economy is still practised by some households. Table 11.6 shows the changes in garden vegetable consumption over the past ten years.

Table 11.6. Percentage Change in Homegrown Vegetable Consumption Patterns by Area, Frequency, and Time

Type of Vegetables/ Frequency	Area 1			Area 2			Area 3		
	Now	5 yrs.	10 yrs	Now	5 yrs.	10 yrs	Now	5 yrs.	10 yrs
Own Potatoes									
Often	34	45	51	20	24	28	45	50	53
Now and then	5	7	7	5	8	7	6	7	6
Never	61	48	42	75	68	65	49	43	41
Own Turnips									
Often	21	30	37	15	19	24	23	28	30
Now and then	6	9	9	4	7	6	7	8	8
Never	73	61	54	81	74	70	70	64	62
Own cabbage									
Often	18	27	34	14	18	22	21	25	28
Now and then	6	8	9	8	7	7	8	9	8
Never	76	65	65	82	75	72	71	66	63
Own beets									
Often	21	30	36	14	16	19	18	21	23
Now and then	7	8	9	4	6	5	7	7	8
Never	72	62	55	83	78	75	75	72	69

Thirty years has seen a big shift in the number of households who produced vegetables for home consumption. Vegetable garden activity for home consumption has become a less frequent activity than thirty years ago when 81% of households grew potatoes, 64% turnips, 60% cabbage, 57% carrots, and 47% beets—all for frequent household consumption. Vegetable gardening has become somewhat marginal in this part of the province and root vegetables are readily available in stores at a reasonable price.

Fruit. In the most recent survey as in the original survey we asked about consumption of the more readily available fruit in local stores such as apples and oranges. Fruit consumption as measured by the items, apples, oranges, and locally harvested wild berries has not changed a great deal in the time fame studied in 2004. Area 2 has the highest frequency of consuming apples and oranges often and Area 3 ranks first for frequent consumption of locally harvested berries. Table 11.7 illustrates the consumption of fruits in the households.

The patterns of consumption of these two types of fruit have not changed appreciably in the thirty years. Apples were consumed often in 75% of households in Area 1 in 1974 and oranges consumed frequently in 67% of households at that time. It could be that consumption of fruit in general has increased because of a wider variety of fruit now available, however fruit consumption in general is lower in the province than the recommended level. This was noted in a recent province-wide study of nutrition done in 1996 (Roebothan 2003).

Gathering of a variety of local berries (including bakeapples—*Rubus chamaemorus*, partridgeberries—*Vaccinium vitis-idaea*, and wild blueberries—*Vaccinium* spp.) has a long history of contribution to household foods in this region of the province (Omohundro 1999). In the early survey (1974) 63% of the households reported frequent use of wild berries, 36%

"now and then" and only 2% never consumed them. This was an increase from the 1960s when a smaller percentage of households reported harvesting wild berries as a readily available source of fruit.

Table 11.7. Percentage Change in Fruit Consumption Patterns by Area, Frequency and Time

Type of Fruit/ Frequency	Area 1			Area 2			Area 3		
	Now	5 yrs.	10 yrs	Now	5 yrs.	10 yrs	Now	5 yrs.	10 yrs
Apples									
Often	64	64	64	79	77	75	65	64	64
Now and then	31	32	32	18	20	22	31	32	32
Never	5	4	4	3	3	3	4	4	4
Oranges									
Often	59	59	59	76	74	72	66	64	64
Now and then	35	35	36	21	23	25	31	32	32
Never	6	6	5	3	3	3	3	4	4
Local Wild Growing Berries									
Often	52	53	52	62	62	63	66	66	67
Now and then	43	43	43	30	31	30	28	28	27
Never	4	4	4	8	7	7	6	6	6

Fast foods. Fast foods, such as fried chicken, hamburgers, and pizza, are currently available just about any where in the world. They are part of the globalization process and the increasing industrialization of food. In many areas of the world as well, eating-out or bringing "take-out" food into the home has become more common to fit in with busy lifestyles. The 1974 survey did not even examine the phenomenon of fast food consumption; it was not really a factor to consider at that time. While restaurants and take-outs are still not a major development in any of the communities studied, the presence of local motels in some communities and resorts near Gros Morne National Park that mainly target tourists or people passing through the communities for business or work, means that there is some availability of these foods. Additionally, people in communities are more mobile and may obtain their fast foods outside the home community.

Our 2004 survey shows that eating in restaurants is a growing trend. In Area 1 there were 74% of households who reported using food from "take-outs" now and then, 73% from Area 3 reporting the same, and 66% of those in Area 2. Fried chicken seems to be preferred over pizza, but that may be a constraint of availability over preference. Of course other types of fast foods, such as French fries or hamburgers, may be used more frequently than these two but these last were not tracked.

Availability and Shopping. Consumption patterns especially those tracked over time give a good deal of information on changes that are taking place in coastal communities around nutritional practices and may also reflect changing food preferences. In order to address food security though we need to examine some of the constraints on these consumption patterns. One is availability of food and what role this may play in household food consumption. A limitation of the 1974 Health Survey was that it did not examine food availability in the local community and what access people had to safe and nutritious foods. We know as indicated in a number of studies (e.g., Felt and Sinclair 1995), that the communities on the Great Northern

Peninsula have changed in the thirty years since that original study was done. No doubt in the mid-1970s choices would have been somewhat limited by transportation of food and type of grocery store in the area. Even now none of the three communities in Area 1 has a supermarket and residents have to depend on travel to a larger centre for grocery purchase. Purchasing in smaller community stores limits choices as our participants indicated. In our 2004 survey we included questions about where people bought the food for their households, how far they had to travel to obtain it, how satisfied they were with commercial food availability, and what were the sources of dissatisfaction as well as improvements they would like to see in access to food.

In Area 1 over 40% of our respondents said they had to travel to a larger community to buy food in a supermarket and an almost equal percentage said their main supply of bought foods came from a local convenience store. The communities are rapidly "aging" as is the province as a whole. Older people are more frequently represented in the groups who depend on small convenience stores and it is perhaps no coincidence that from the 1996 nutritional survey older people especially older women were most at risk nutritionally (Roebothan 2003).

Satisfaction with commercial food availability. A number of people in the communities in each area expressed dissatisfaction with availability of foods locally. While there was difference in satisfaction with what food was available in the local stores among the three research areas, residents in Area 1were most dissatisfied (37%). Their most frequent source of dissatisfaction (29%) was a lack of access to sufficient fresh produce, mainly fruits and vegetables. This was followed by the complaint of "not much variety" (14%), and "not much fresh meat" (13%). Other areas of dissatisfaction mentioned by a few participants were that food was costly or people were advised to follow a certain diet for medical reasons and the food they required was not readily available.

Given the sources of food availability in local communities and lack of access to transportation for some residents to obtain food in larger centres, it is not surprising that local food supply could be improved by having a greater supply of fresh produce including fresh meat, and having a greater selection of food and lower prices. A few felt that having a local supermarket (that hopefully would address these areas of dissatisfaction) would be a way to improve food supply in the community. Not having these issues addressed contributes to the sense of marginality that people in these communities experience.

CONCLUSION

Some of the food consumption patterns have changed greatly over the past thirty years in the set of communities where we have longitudinal data (e.g., Area 1) and indeed even in the past ten years there have been a few notable changes across the region. The surveys revealed a number of findings related to changing food consumption patterns. The first is that these patterns vary from region to region even within a defined geographic area and they are affected by a number of socio-economic factors. Second, food consumption patterns despite the wide array of factors that affect them are relatively stable over time, so it generally takes a relatively long period of time to effect change. The findings from the research suggest a number of broader issues relating to food consumption patterns in a global context. The first is to what extent do the findings mirror what is happening in other parts of the world? The second is what can these patterns tell us about food security and factors that affect them?

The consumption of red meat, such as beef, has shown marked decline in the United States, where it went from 79% in 1970 to 62% in 2000 (Haley 2001; Regmi 2001). In around the same period in Area 1 commercial beef consumption went from 93% to 61% if we look at any consumption, i.e., "often" or "now and then" of beef in these households. Other changes in United States per capita consumption over the same period have been decreases in coffee, meat, and eggs, and increases in alcoholic beverages, fruits and vegetables, fats and oils, fish, grains and cereals, poultry, and cheese. While we were not able to look at all these foodstuffs in our study, we did notice an increase in use of certain fats and oils. We did not see the same increase at least in the use of fruits included in the survey. In fact there was a decrease in consumption of apples and oranges in Area 1 over the thirty-year period. Other fruits (i.e., bananas) could be consumed more frequently, because at present most of the communities have a greater choice in the fruits that are available. However, apples and oranges are the leading fresh fruits available in North American markets (Pollack 2001).

Other studies in Canada have indicated changes in consumption of certain foods. Gray-Donald *et al.* (2000) noted that levels of fat consumed are decreasing over time in Canadian provinces. We did observe a change in fats consumed as well as a decrease in the use of selected fats such as salt pork fat and lard. Our study can also help us to understand food security, culture, resilience, and health in these coastal communities where these surveys have taken place, and the factors that may be driving the changes we observed. An important question to revisit is "Are there elements of food insecurity in this region?"

Food security. If we look at the definition of food security that came out of the World Food Summit (1996) and cited in the introductory chapter to this book and look at the various elements contained in that definition, physical and economic resources to meet "dietary needs and food preferences" are threatened. While some residents do have the economic resources to afford nutritious food, the statistics profiling the communities tell us that there are vulnerable groups. Social assistance incidence is fairly high and this incidence does not pick up those with few economic resources who because of part-time or seasonal work are able to remain off social assistance payments. A recent discussion paper on *The Cost of Eating in Newfoundland and Labrador—2003* (Dietitians of Newfoundland and Labrador et al. 2004) identifies some vulnerable groups and some of these groups are overrepresented in the study area. They suggested the elderly are vulnerable and there is a high percentage of elderly people in these three communities. In fact the highest percentages of seniors in Canada who receive the Guaranteed Income Supplement (Old Age Pension) live in Newfoundland and Labrador. Furthermore this report suggests that the highest cost of eating for those who are seventy-five years and older is on the Great Northern Peninsula. Low-income earners, or those on fixed incomes, are also a vulnerable group (Gehlar and Coyle 2001) and given the levels of income in the communities, this would be a factor. The importance of economic factors is recognized in the definition of food security and one has only to examine the literature on poverty and nutrition to recognize the impact that low and poor incomes have on food security at the individual and household level (Hamelin et al. 1999).

Food security is further threatened because households do not have easy physical access to the food they prefer as is evidenced by their sources of dissatisfaction with local food supplies. A fact that cannot be discounted in this study is that we are dealing with a largely rural area and rural areas, despite the effects of globalization, have more limited choice in supply and lower demand than larger urban areas (Regmi and Dyck 2001). As out-migration and lower birth rates in these rural areas continue to result in population decline and aging communities, the effects may further serve to threaten food security. Aging as a variable in food consumption patterns cannot be overlooked.

While advances in transportation technology have enabled the delivery of perishable goods to further reaches of the world at lower prices and reasonable quality there are some limitations (Coyle et al. 2001). It is usually larger markets that are targeted leaving some rural areas still at a disadvantage despite these advances. The transportation factor that is important to residents in our communities is generally availability of personal transportation that allows people to travel to larger centres for food. Older people and the poor have less access to this transportation.

Culture. The definition of food security explicitly acknowledges the cultural aspects of food and the importance of these aspects to food security. Our work with First Nations people as illustrated in this volume by Ommer *et al.* and Tirone *et al.* (Chapters 8 and 10) speaks to the importance of culture and food security. In addition certain environmental factors or changes in the environment may have a direct or indirect effect on food security that interacts with culture. Montevecchi *et al.* (Chapter 7 this volume) nicely illustrate how the environment and patterns of marine bird hunting, which provided locally obtained food, are linked. Looking at the place of fish in the food consumption in the communities studied, and despite the changes that have taken place in the fishery, fish (cod, salmon, and herring) are still being consumed. Other food that is part of the Newfoundland diet, such as seal, continues to be eaten as well. People in the communities prepare seal, salmon, and moose as they have for a number of years by bottling these products.

Fresh homogenized milk is at least somewhat replacing canned evaporated milk, and fat consumption patterns have changed with cooking oil replacing lard and salt fat pork. The latter two are more associated with what we think of as the "traditional" Newfoundland diet although as Amilien (2001) indicated this concept can have different meanings. Other foods in the category of traditional foods in the province, salt fish and salt meat, have also been declining in frequency of consumption. While these changes may result in positive health benefits, they reflect an erosion of some aspects of cultural identity in food.

Resilience. Several strategies for production for household food consumption speak to resilience in these communities. One of these is the use of local berries to supplement the diet. Berries are used to make jams and jellies and at the same time probably contribute significantly to levels of vitamin C and other important nutrients in the diet. Some of the households rely solely on these sources and do not purchase them commercially. Another strategy is the use of home gardens. Some residents are still growing potatoes, turnips, carrots, and cabbage. The third strategy is hunting, particularly of moose. Households continue to depend on moose as a readily available supply of protein.

Game, in the form of moose, shows high consumption rates and this is not surprising given the high availability because of population increase of moose in the area and the importance of hunting as an activity in rural Newfoundland and Labrador. Moose meat serves as a good low fat source of protein. It has been promoted by the Department of Health in the province as a good source of protein and is widely used in households in the study area. This is positive for the communities as it is readily available, and capitalizes on the hunting skills of the residents. Moose hunting is good for social cohesion and allows for sharing within the community. Moose meat in Newfoundland and Labrador is generally considered safe from major contaminants, such as cadmium, as long as kidneys and liver are not consumed (Gamberg 1997).

Health. While it is important not to equate changes in dietary practices with immediate health effects, there are positive and negative implications for health relating to the changes in food consumption. Our findings suggest there is less reliance on consumption of some of the foods that are thought to have a negative impact on health, e.g., salt beef, salt fat pork, and salt cod. Some of this change would be in response to campaigns to decrease "salt" in the Newfoundland and Labrador diet because of health concerns. It should be noted that the communities as part of the Health Survey in 1974 were part of a hypertension study (Marshall 1975), and hypertension is a disease condition associated with high salt intakes.

Another diet change that would be expected to have positive health effects is the change in fat consumption. Our findings show that there has been a shift in the consumption patterns of the fats we investigated. Vegetable oil is replacing lard for cooking and homogenized milk is to some extent replacing evaporated, condensed milk. Heart disease is very common in the province and one of the health promoting and disease preventing strategies used in Newfoundland and Labrador, as elsewhere, is to have people reduce their consumption of fat. During the interim period between the two surveys, Canada's food guide was changed and in 1982 one of the changes was a recommended decrease in fat intake (Canada, Health 2002). This information would have gradually filtered down to the local level and would have some influence on local diets.

Fruit and vegetable consumption is important because it is believed to be an indicator of healthy eating and it is encouraged as a strategy for disease prevention (CIHI 2004). In particular, eating high proportions of fruits and vegetables is consumption of fresh fruit is thought to be protective against heart disease and while the evidence is conflicting, some advocate the consumption of these food groups to prevent cancer. In nationals surveys the consumption of fruit by Canadians is lower than that recommended by Canada Food Guide. Despite the promotion of consumption of fresh fruit and vegetables both in the United States and Canada our results are in keeping with recent nutritional studies (Morland et al. 2002; Roebothan 2003; Krebs-Smith et al. 1995). The consumption of fruits in the study region needs further attention; promoting the use of local wild berries may be a good strategy for enhancing health in these communities.

At the individual level decisions around food choice are complex (Birch 1999; Connors et al. 2001; Dibsdall et al. 2002; Drewnowski 1997). There are as well a host of factors beyond the individual level that will affect food choice and it is important to focus on some of these factors (Beaudry et al. 2004). We need to work to improve diets to in rural communities and reducing some of the factors that contribute to poorer nutritional practices. Strategies may include general education, lobbying food suppliers to improve quality and selection of food, or nutrition programs in the school. It is also important not to ignore larger issues that influence household food purchase and thus consumption, such as adequate household incomes, legislation of food, or advertising of certain products. We must be aware as Keane (1992) warned of the politics involved in factors that affect food security. Most of the information that we get about food comes from commercial interests and governments are generally reluctant to become involved. Likewise, the same interests control food availability at the local level. Ensuring food security at the household and community level requires a multifaceted approach.

In: Resetting the Kitchen Table
Editors: C. C. Parrish et al., pp. 177-190

ISBN 1-60021-236-0

Chapter 12

HISTORICAL PERSPECTIVES ON NUTRITION AND FOOD SECURITY IN NEWFOUNDLAND AND LABRADOR

Linda Kealey

Department of History, University of New Brunswick, Canada

ABSTRACT

Food security is a relatively recent concept, but nutrition, diet, and dietary disease have captured attention over the past several centuries. This chapter focuses on diet and dietary diseases in Newfoundland and Labrador as one way of getting at the historical record of human health. Recognizing the importance of a good diet, twentieth century public health advocates and specialists in nutrition, for example, emphasized a varied diet, including "protective foods" that encouraged health or mitigated disease. Debates about the reasons for the inadequacy of working-class diets stemmed back to the turn of the century in the United Kingdom and the large percentage of unfit recruits for service during World War I drew attention to the poor health and poor diets of many working-class men in North America as well as the United Kingdom. By the 1920s, health experts began to understand the importance of vitamins in a healthy diet and were able to connect the lack of specific vitamins to particular diseases. The Great Depression of the 1930s also called attention to poverty and malnutrition with so many people on relief. In Newfoundland and Labrador, concern with diet and dietary diseases first appeared just prior to WWI; subsequently a dozen and a half studies were carried out to explore the effects of limited diet on the population. Indeed the area became a laboratory for scientists and medical doctors because of its isolation, homogeneous settler population, and reputation for poverty.

The analytical framework of the chapter employs the tools of social, women's, and medical history as well as feminist theory. It suggests that we need to understand not only the context and world views of male medical scientists and doctors but also the central roles played by women as household managers providing for their families and as nutritionists in a newly formed profession. While male experts directed most dietary studies, some women were also involved as researchers; others were employed by or volunteered with voluntary groups such as the Grenfell Mission located in northern Newfoundland and founded in 1892 by Dr. Wilfred Grenfell of the Royal National

Mission to Deep Sea Fishermen. Historically a fishing station within the British Empire, Newfoundland became a settled colony with a family-run fishery by the early nineteenth century despite British attempts to prevent settlement. Thus the imperial context also helped shape the direction of health care and the approach to dietary deficiency diseases on the island until the 1940s when Newfoundland began to "modernize" and became a part of Canada in 1949. In the post World War II period dietary deficiency problems continued to preoccupy health officials; they adopted many of the recommendations of the influential Cuthbertson study (1947), which built on previous studies conducted in the interwar and wartime years. Government increasingly assumed the lead role in establishing policies, programs, and institutional bodies that would improve health. Ironically, while women were often blamed for dietary problems, they were also the trained experts who assumed responsibility for providing the necessary expertise, albeit under the direction of doctors in the government bureaucracy.

INTRODUCTION

While "food security" is a relatively recent concept, concern for adequate nutrition has captured attention over the past several centuries. Recognizing the importance of a good diet, twentieth century public health advocates and specialists in nutrition, for example, emphasized a varied diet, including "protective foods" that encouraged health or mitigated disease. Debates about the reasons for the inadequacy of working-class diets stemmed back to the turn of the century in the United Kingdom (Smith and Nicolson 1997) and the large percentage of unfit recruits for service during World War I drew attention to the poor health and poor diets of many working-class men in North America as well as the United Kingdom. By the 1920s health experts began to understand the importance of vitamins in a healthy diet and were able to connect the lack of specific vitamins to particular diseases (Guggenheim 1981, 155–83). The Great Depression of the 1930s also called attention to poverty and malnutrition with so many people on relief. In Newfoundland and Labrador, the focus of this chapter, concern with diet and dietary diseases first appeared just prior to WWI; subsequently eighteen studies were carried out to explore the effects of limited diet on the population. Indeed the area became a laboratory for scientists and medical doctors because of its isolation, homogeneous settler population, and reputation for poverty.

This chapter focuses on diet and dietary diseases as one way of getting at the historical record of human health. The analytical framework employs the tools of social, women's and medical history as well as feminist theory. It suggests that we need to understand not only the context and world views of male medical scientists and doctors but also the central roles played by women as household managers providing for their families and as nutritionists in a newly formed profession. While male experts directed most dietary studies, some women were also involved as researchers; others were employed by or volunteered with voluntary groups such as the Grenfell Mission located in northern Newfoundland and founded in 1892 by Dr. Wilfred Grenfell of the Royal National Mission to Deep Sea Fishermen. Historically a fishing station within the British Empire, Newfoundland became a settled British colony with a family-run fishery by the early nineteenth century. Thus the imperial context also helped shape the direction of health care and the approach to dietary deficiency diseases on the island until the 1940s when Newfoundland began to "modernize" and became a part of Canada in 1949. In the post World War II period dietary deficiency problems continued to preoccupy health officials who adopted many of the recommendations of the influential Cuthbertson study (1947) which built on previous studies conducted in the interwar and wartime years.

COLONIAL CONTEXTS

Initially valued for its fishing grounds and the salt fish trade, by the twentieth century Newfoundland had become a Dominion within the empire with a population of just over 290,000 in 1935 (Newfoundland, Public Health and Welfare 1935). Although the 1920s were economically difficult, the Depression hit the island hard with one-quarter of the population on the dole, that is, public relief, in the winter of 1932–33.With a rising debt and threat of bankruptcy on the horizon, the Amulree Commission recommended that Britain accept financial responsibility for the Dominion. In return the government of Newfoundland suspended parliamentary self-government and agreed to replace it with a Commission of Government featuring a Governor and six Commissioners, each one of whom was responsible for a particular department (Neary 1988). Sir John Puddester headed up the Department of Public Health and Welfare with the assistance of Dr. H. M. Mosdell, former chair of the royal commission on public health that reported in 1930. From this report's recommendations came the 1931 Health and Public Welfare Act (22 Geo. V, c.12) which consolidated public health legislation and mandated significant changes to the system. Most of these changes were not implemented until the Commission of Government acted on them during its 1934–49 mandate. Some of the reforms undertaken during the Commission period included the construction of cottage hospitals (based on the Scottish model), the creation of a district nursing scheme, travelling medical clinics, a tuberculosis sanitarium and a midwifery training program, to name a few. In general these reforms stressed taking services to the outports (small fishing villages scattered around the coast) and distant settlements where medical services were scarcest.

By the time of Commission of Government reforms northern Newfoundland and coastal Labrador had been served for over forty years by the explicitly Christian medical mission begun by Wilfred Grenfell. Set up to serve the medical and religious needs of the fishing population along the coast, the mission had evolved by the 1930s into a large and complex organization of hospitals, clinics, nursing stations, orphanages, schools, industrial centres, agricultural stations, hospital ships, cooperative stores, a lumber mill, and a ship's repair station. Staffed by paid and volunteer professionals as well as local tradesmen and service workers, turnover was a constant challenge particularly among doctors and nurses. The organization sought to attract professionals with medical qualifications, as well as midwifery and dentistry, but they also sought those with a strong sense of Christian stewardship. Even after government reorganization of health care, the Grenfell Mission continued to provide healthcare to these northern areas, albeit with increasing reliance on government funding which gradually replaced the fund-raising activities of the International Grenfell Association and its branches in Canada, England, and the United States. As the archival and oral sources indicate, the Mission retained its colonial and paternalistic attitudes well into the 1970s (Rompkey 1991).

The imperial authorities in the 1930s were increasingly aware of health and its relationship to diet. Proper nutrition and women's role in providing healthy food attracted the attention of the Secretary of State for the Colonies, J. H. Thomas, in 1936 when he requested information on nutrition from many areas of the empire, especially Africa and the Caribbean. Published in 1939 the report stated: "We have no doubt that improved nutrition will bring very great benefit to the Colonial Empire . . . More important than the effect of malnutrition in directly producing disease is its effect in producing general ill health and lowered resistance to other diseases, inefficiency of labour in industry and agriculture, maternal and infantile mortality and a general lack of well-being." Expenditure on nutrition would lead to "greater efficiency in production and less waste of human life and effort." Malnutrition, the authors wrote, was caused by low standards of living, ignorance, and prejudice as well as disease. What could be done? The authors recommended subsidization of food costs, cheap freight rates, the elimination of duties, and increased home production. The report urged that women be targeted since they were key in many cultures in deciding what to grow; they also reaped, stored, cooked, and distributed the products as well as fed the children and did the marketing. The report also recommended bringing more local women into government welfare work and specified that "most forms of propaganda should be directed primarily to the women and more particularly to the mothers," who would ensure the transmission of information from generation to generation. These gendered themes of women's responsibility for welfare work generally and for nutritious food in particular were echoed in discussions and policies surrounding diet and health in Newfoundland and Labrador (Great Britain 1939, 12, 39, 122).

EXPERTS TACKLE DIETARY PROBLEMS IN NEWFOUNDLAND AND LABRADOR

As Wilfred Grenfell was quick to discover, many of the health problems encountered in northern Newfoundland and coastal Labrador stemmed from the restricted diet available to residents whose fortunes were tied to the fishery and the merchant who furnished supplies on credit. Using this truck system, fishers obtained supplies on credit and paid back the merchant at the end of the fishing season though in bad seasons, the debt might not be paid. Typically, fishing families purchased six or more months of supplies to last the winter when shipping became difficult as harbours froze up. Fishers bought white flour, tea, molasses, salt meats, dried peas, beans, fruit, oleomargarine, and tinned milk and supplemented their diet with wild meat, fish, root crops, and berries available locally. By spring many families ran out of supplies and lived on salted meat or fish, bread and tea. The result was malnutrition and dietary deficiency diseases such as beri-beri, night-blindness, scurvy, and rickets. In the early twentieth century knowledge of the causes and remedies of these diseases was limited. Beri-beri, for example, was associated with diets based on polished rice but it was also identified by Dr. John Mason Little of the Grenfell Mission as being associated with the monotonous diet of fishers heavily dependent on white flour (Little 1912, 1914). Eventually beri-beri was understood as resulting from vitamin B deficiency but the disease remained a concern of researchers and government into the 1940s. During the Depression the government imported whole-wheat flour to be used for relief. Despite efforts to persuade dole recipients to accept the flour as a temporary solution to the beri-beri problem, those on the dole objected to the flour as inferior in quality and stigmatizing (Overton 1998).

The early studies by Little also helped to increase interest in the connection between diet and disease. During 1919–20, Dr. Vivia Appleton studied communities on both sides of the Straits of Belle Isle finding that the Labrador side exhibited a different pattern than the Newfoundland shore. Deficiency diseases occurred with much more frequency on the Newfoundland side where she found over one hundred cases of beri-beri and equal incidents of xerophthalmia. Infant mortality was also higher. She concluded that "the people in both regions are in a 'twilight zone' where very slight changes in the diet may cause deficiency diseases." The scarcity of fresh fish after August and the lack of canned milk and vegetables accounted for the higher incidence of some diseases on the Newfoundland side. Despite the presence of more cows on the Newfoundland shore, the lack of sufficient pasturage hindered milk production and cream was most often made into butter. While Labrador had few cows, the fishing season was longer providing fresh fish into October and there was evidence of more use of canned milk and vegetables. Overall, Appleton noted that the children were smaller in size and undernourished with poor teeth. She also noted that young women in particular suffered from amenorrhea (absence of menstruation) as a result of undernourishment and that mothers trying to breastfeed their children were not able to do so successfully if their diet was insufficient (Appleton 1921).

In the late 1920s at the request of the Grenfell Mission, another woman, Dr. Helen Mitchell of Battle Creek, Michigan along with her assistant Marjory Vaughn, visited a dozen communities in northern Newfoundland and Labrador. Bringing with her various food products donated by large companies and often accompanied by the local nurse or industrial worker, Mitchell offered the food to the families who participated in the survey hoping to demonstrate the benefits of orange juice, bran, fresh vegetables, and canned milk. Calculating the nutritive value, Mitchell concluded that inhabitants received enough calories but lacked a balanced diet (Mitchell 1930). Under Mitchell's supervision, Vaughn continued the project in 1930 in the Northern Peninsula community of Flowers Cove. Working from their observations of the limited diet and the need to improve the intake of calcium and vitamins, Vaughn undertook extensive educational work to try to demonstrate the need and the possibility of increasing local produce through gardens. She also organized "Health Week" in the community with baby shows, health lectures, cooking demonstrations, and exhibitions of foods that prevented common diseases. An agricultural show encouraged people to bring in their products; in the fall, a school lunch program began for sixty children, undoubtedly the first of its kind on the island. Vaughn's report indicated her satisfaction with the program noting that some children gained weight and looked "rosy." In the spring she visited the nearby fishing settlements urging people to obtain seeds from the Mission store and to plant root crops and other vegetables. While this resulted in the expansion of gardens that year, judging from the "sustained tradition of making vegetable gardens," observed by Dr. Gordon Johnson, the effects may have been longer term (Vaughn and Mitchell 1933; Johnson, n.d.).

At the same time as Mitchell and Vaughn were working in Flowers Cove, Irish-born physician Dr. Wallace R. Aykroyd began to investigate deficiency diseases. As early as 1928 he had published a paper on night-blindness which occurred in an estimated one-quarter of the population in Newfoundland. His work clearly separated night-blindness from other diseases (such as beri-beri) and tied it to Vitamin A deficiency (Aykroyd 1928). In another study by Aykroyd, examination of the records of the St. Anthony Hospital revealed that up to 12% of admissions in the April–June months of the years 1912 to 1928 were due to beri-beri. Those families with beri-beri consumed few vegetables and little fresh meat and like

Appleton he noted that the disease was less common in Labrador because of higher consumption of game. Unlike other researchers such as Appleton, however, Aykroyd identified beri-beri as a disease of poverty and specifically tied its occurrence to the success or failure of the previous fishing season. Furthermore its debilitating effects often prevented a person from going fishing the next season particularly if their diet had been set by dole rations. As Overton noted, Aykroyd thus felt that education had little to do with eradicating beri-beri. According to Aykroyd, the efforts of the Grenfell Mission to introduce whole-wheat flour had met with very limited success and the government would be better off recognizing the disease as the result of poverty and instituting economic reforms (Aykroyd 1930; Overton 1998, 10–12).

Nutritional deficiencies captured the attention of a number of researchers and research teams during World War II and the years immediately following. Dr. Robert Dove, resident physician at the Bonne Bay Cottage Hospital in Norris Point on the West Coast, and several others published a study of subclinical dietary deficiency diseases in his district. The work was carried out in the winter of 1941–42 and its purpose was to assess "the effects of prolonged deprivation of normal dietary intake of vitamins" on the population of these West Coast communities. In describing the setting, the economy, and the population, the authors confirmed earlier observations relating to the scarce milk supply and noted the absence of transportation, storage, and refrigeration facilities as well as low levels of education. The study noted that "The fact that vegetables and berries are listed as available does not mean necessarily that these are utilized, for many persons restrict their diets through ignorance, prejudice and faulty preparation of foods. Dairy products are extremely difficult to obtain and even where cattle are kept there is apt to be a traditional distaste for milk which is used only sparingly in tea." In addition to noting vitamin B deficiencies, the study pointed out that 70% of the population had normal ranges of blood vitamin C for only two months of the year. Thus while patients checked in the fall showed satisfactory levels of the vitamin, by spring their levels were deficient and clinical signs included dental caries, gingivitis, and hyperkeratosis. Vitamin C deficiency, the study noted, was particularly likely for pregnant women and their infants and the authors cautioned that the incidence of infantile scurvy was probably higher than previously thought, possibly contributing to high infant mortality in the whole country (McDevitt et al. 1944; Dove 1943).

This study and one by Dr. Olds (1943) on vitamin C deficiency in the Twillingate area appear to have prompted the Newfoundland Medical Association to set up a Nutrition Council in 1943. The Council successfully approached the Commissioner for Public Health and Welfare in 1944 asking for a major survey of the Avalon Peninsula and the Fortune Bay area. A group of experts from Great Britain, Canada, and the United States that included Dr. B. S. Platt, the director of Nutritional Services in the Colonial Office, conducted the survey in eight communities in August and submitted a preliminary report by the end of the month with a fuller study published in the *Canadian Medical Association Journal* in 1945. Even in the preliminary version, the group noted a high prevalence of deficiency disease and blamed it on the "dietary habits of the people" (Nutritional Survey of East Coast Newfoundland 1944). In addition to recommending food fortification, an increase in the supply of milk, school lunch programs, and vigorous educational programs, the study advocated prevention rather than cure. In particular the authors recommended that the government set up a nutritional council, noting that "most governments in British colonial territories in the past eight years have had nutrition committees . . . " They also added that a recent UN conference recommended the creation of national nutrition councils; ideally these bodies would cooperate with a permanent world organization set up to deal with such problems. Specialists in nutrition, agriculture, forestry, and fisheries would be sent to assist in setting up more complete programs to deal with what the authors concluded was in large part an economic problem. Later in 1944, Puddester took the initiative and set up an official Nutrition Council for Newfoundland that was separate from the Council of the Newfoundland Medical Association (Adamson et al. 1945; Newfoundland Nutrition Council 1944).

Adamson *et al.*'s detailed study was accompanied by startling photographs of individuals with various conditions attributed to the diet. Photos included shots of skin conditions, mouth, gum and teeth problems, and skeletal deformities (Adamson et al. 1945). The findings combined with sensationalistic illustrations drew the attention of other medical journals as well as popular magazines. While the *Lancet* queried the role of salting fish and meat in potentially decreasing nutrients, the popular *Magazine Digest* of October 1945 compared Newfoundland to the Virgin Islands "where the white 'ruling classes' have steadily opposed political liberty and social and economic betterment for the native population. They contend that colored folk in tropical climates are' just naturally lazy' as compared to the more energetic white races in colder climates . . . " Thus, the writer continued, poor nutrition was rejected as a possible reason why native populations preferred to work only the minimum number of days needed to provide the bare necessities. Responding to a United States Army engineers' report that Newfoundlanders only worked enough to get by, the author concluded that "All that the Newfoundlanders and the Virgin Islanders have in common is this alleged indifference towards hard work—and a diet well below the level required to maintain health and strength." The article furthermore claimed that the study showed that Newfoundlanders were literally committing suicide by their dietary habits and the story featured the alarmist headline "Why three out of four 'hardy' Newfoundlanders will never live to be 40" (*Lancet* 1945; Henry 1945). In contrast to these sensational views, a critique in *Nutrition Reviews* the same year noted that there was no study of the availability of medical and dental services which might account for the high death rates from tuberculosis among infants and the poor quality of dental health. Without such information it would be difficult to evaluate the findings (*Nutrition Reviews* 1945).

At almost the same time in 1944, a smaller survey occurred in Norris Point on the West Coast of the island conducted by Metcoff and others, including Dr. R. F. Dove and his wife, Margaret. This study targeted those most vulnerable in the population, namely, women and children. The imminent introduction of enriched flour in the diet also encouraged a baseline study so that the effects could be followed. The authors noted at the beginning that they had difficulty taking complete histories of their subjects because of their reticence and shyness, which was complicated by "certain colloquialisms of speech." Like the Adamson study, a number of measures were used including nutritional and medical histories, a physical examination, photographs, and laboratory tests for sugar, albumin, cellular material, hemoglobin, and plasma protein. The authors also commented on the general attitudes and reactions of the women and children who comprised the majority of the study's subjects (97/113; Metcoff et al. 1945). They noted that the women and children were "extremely shy and often apathetic. The children, especially, lacked spontaneity and were inactive and quiet. There was no enthusiasm or exuberance of spirit and little inquisitiveness was manifest during the examination procedure . . . and in adults a lack of initiative with an attitude of acceptance of difficulties was the rule . . ." Noting that Mitchell had made similar observations, the authors conceded that lack of organized play might have contributed to the situation; there was no recognition of the possible effect of strangers in the community, however (Metcoff et al. 1945, 475).

Perhaps most interestingly, the draft report sent to officials at the Department of Public Health and Welfare aroused some objections to its language. The Commissioner, J. C. Puddester, pencilled a note on the bottom of the cover letter written by Harvard's Dr. Frederick J. Stare to Dr. James McGrath, Assistant Director of Medical Services, which said: "I cannot see the necessity of further publicity of our 'abject poverty' etc. that they refer to. Anyway we have enough information now to enable us to think about a proper remedy. It is time now that the American and Canadians who are so anxious about us to turn their attention to some of their own back yards." In a subsequent letter from McGrath to Stare, the former asked Stare to remove descriptions of the area as possessing "an inhospitable depressing climate" and a description of the children as undersized in height and weight compared to healthy white American children. McGrath noted that this contradicted the Adamson report's findings and that the comparison group ought to be underprivileged American children. Noting that newspaper reports tended to dwell on the unfavourable, he stressed that they wanted "to avoid any suggestion that these are exclusively characteristic of Newfoundland." While Stare agreed to most of the changes he objected to removing the statement regarding the children's heights and weights since he believed that data supported the statement that "the children appear to be undersized" (Norris Point Study 1945).

Again, the study revealed that the incidence of deficiency was highest among the women especially in riboflavin, iron, and Vitamin A and many mothers had combinations of these deficiencies. In conclusion, the authors noted that "An isolated community such as Norris Point is an ideal laboratory for investigation of the value of various therapeutic procedures" (Metcoff et al. 1945, 487). A resurvey of Norris Point in 1948 by Goldsmith et al. found notable improvement in relation to Vitamin B, a finding consistent with the use of enriched flour. While there were signs of improvement relating to Vitamin A, there had been no improvement with levels of Vitamin C. The relationship between nutrition and pregnancy provided the topic for another paper written by Grace Goldsmith based on the Norris Point data. She found that recently pregnant women were most at risk of Vitamin B complex deficiency and thus the enrichment of flour with these vitamins benefitted this group of the population most. In addition there was a reduction in infant mortality and still birth rates (Goldsmith et al. 1950).

In the post-Confederation period after 1949, far fewer studies were conducted as the new provincial government dealt with integration into the Canadian system (Goldsmith et al. 1950; Goldsmith 1950; Scobie, Burke, and Stuart, 1949). In the late 1950s, however, three medical officers in St. John's found cause for concern in the increasing incidence of infantile scurvy. They identified seventy-seven cases in Newfoundland in 1959, most of them occurring between the ages of six to twelve months; they estimated that the incidence of infantile scurvy in the zero to two age group would be 250 per 100,000, nearly twice the rate of poliomyelitis which was epidemic in that same year. Despite the availability of free concentrated orange juice, the problem persisted causing the doctors to inquire into the ascorbic acid content and storage of the juice; since most families did not own refrigerators, the juice lost its ascorbic acid content when stored at room temperature. They also used questionnaires, interviews, and telephone conversations to find out why the vitamin C supplement had not been used. Lack of knowledge about the need for vitamin C occurred in three-quarters of the cases. The doctors also found that none of the forty mothers they were able to contact were breast-feeding which might have solved the deficiency. While breast-feeding was more common before 1949 and some used fresh cow's or goat's milk in infant feeding, changes had occurred since Confederation, including more use of cash: "Today, practically all infant feeding depends upon the use of evaporated milk which in depression days was a luxury. There is an undue dependence upon food from cans, packet cereals and other commercial preparations." In addition the idea persisted that "breast feeding may predispose to tuberculosis in mother and child" in some areas and many regarded the practice as a sign of poverty. The authors concluded that the best solution was to fortify evaporated milk with vitamin C just as had been done with vitamin D. By 1964, this became the practice (Severs, William, and Davies 1961, 217–218).

These studies conducted on various populations and locations identified common and sometimes serious deficiencies in the diets of inhabitants particularly in isolated communities where fresh food and milk were not available, depending on season, or were too expensive. While an occasional study fastened on poverty as the underlying problem requiring a solution, most advocated changes in diet, more gardens, improved storage and transport systems, more education, and eventually the fortification of certain foods. The effects on women and children became a focus of a number of these studies. Newfoundland and Labrador over the course of the century had served as a kind of laboratory for medical investigators intrigued by the persistence of dietary diseases that were associated with underdeveloped "traditional" societies. In the process officials attempted to change the negative image of Newfoundland and Labrador as a place of poverty and malnutrition by embracing modernization not only of the fishery but also of nutrition and health.

GOVERNMENT RESPONSES AND POLICY INITIATIVES

The Commission of Government era, 1934–49, witnessed major reforms in the health care system, much of it initiated by government officials who were committed to providing health care to remote areas. While Dr. John Olds of Twillingate developed his own form of a medical pre-payment plan when the Department of Health claimed it could not increase his budget, the Deputy Minister of Health, Dr. H. M. Mosdell visited Scotland to observe the cottage hospital system himself, as the government launched a cottage hospital scheme in Newfoundland (Saunders 1994, 97–107). The early dietary studies of Appleton, Mitchell and Vaughn and Aykroyd suggested the widespread nature of the problem as well as steps that might be taken to improve the situation such as school lunch programs, the encouragement of gardening and an education program. The flurry of studies undertaken in the 1940s during the war also produced new policy directions. The two major surveys of 1944 by Adamson and Metcoff, for example helped to establish the Newfoundland Nutrition Council and encouraged a further major review undertaken in 1945 by Dr. D. P. Cuthbertson of the British Medical Research Council (Adamson et al. 1945; Metcoff et al. 1945; Cuthbertson 1947). The Cuthbertson report seemed to galvanize the bureaucracy in a way that previous reports and recommendations did not. Influenced no doubt by the negative publicity surrounding the earlier studies, government officials were concerned to correct the impression that Newfoundland was a land of hunger and want. Building on earlier studies as well as the efforts of individual doctors and the Newfoundland Medical Association, Cuthbertson's report and recommendations were the focus of several meetings of the Nutrition Council in mid-1945 that were attended by representatives of four government departments. His advice included the enrichment of margarine and flour with vitamins, incentives to increase cod liver oil production, wider milk distribution, more encouragement of breastfeeding, restocking the island with moose and caribou, more consumption of fruit, limits on the export of bakeapples, and an aggressive educational program in nutrition. Dr. Cuthbertson highly valued education to change cooking and eating practices; although dietary habits might represent "the unwritten cumulative experience of countless generations of men and women," information on how to cook food properly without losing the nutrients "cannot permeate rapidly unless active steps are taken through existing and new educational channels." Furthermore, he urged every government and voluntary agency to become involved in the campaign to spread technical information concerning gardening and preserving produce as well as the health benefits of fish and cod livers (Cuthbertson 1945).

While the government had already initiated a number of programs to combat dietary problems, Cuthbertson's report and recommendations proposed a multi-pronged and coordinated attack. By 1945 flour enrichment with several types of Vitamin B and iron and the fortification of margarine with Vitamin A was in place. Subsequent to the Cuthbertson report the fortification of margarine increased the Vitamin A content. By 1947 expectant and nursing mothers and infants received orange juice and bone-meal provided additional calcium enrichment in flour. That same year cod liver oil and cocoa milk powder were distributed to the schools (Newfoundland Department of Health, n.d.). Thus government put into place two key interventions recommended by Cuthbertson: enrichment of foodstuffs with vitamins and minerals and the direct provision of supplements to vulnerable populations of women and children. A third key element was the provision of education so that eating habits would change over the long term.

WOMEN'S ROLES IN CHANGING DIETARY PRACTICES

Ultimately the responsibility for good health, proper nutrition, and changing dietary habits fell to women whether as wives and mothers or as outside experts. Although the former were often chastised for their ignorance of proper cooking and preserving methods, the latter group represented a new profession of educated women with expertise in nutrition. While the government first recognized the potential assistance that could be rendered by someone trained in home economics and nutrition in 1943 with the part time work of Edna Baird of Memorial College in St. John's, the Grenfell Mission had already experimented with dietary reform since the early twentieth century. Experiments in agriculture and animal husbandry largely failed so that by the 1910s nurses and teachers were encouraged to teach local women and young girls proper methods of cooking and hygiene as well as promoting gardening.

By 1920 the Mission had decided to welcome a number of American women who had been professionally trained in dietary methods. Approximately thirty nutrition workers, most of them students and volunteers from the United States and teachers trained at the mission spent summers in the 1920s carrying out nutrition work which included health classes for local women and children, school lunch programs and gardens, baby and dental hygiene clinics. Although some of these women were graduates of female professions, such as nursing or social work, additional training in home economics was essential for certification in nutrition work. A nutrition worker was described as "one who worked with the physician on the nutrition of children either in nutrition classes, or the homes, or both" (New York Nutrition Council 1921, 493). In the early 1900s home economists controlled dietetic training for homemakers, career-destined women, and medical students. In the post World War I era, they sought to create a new profession based on their core dietetics subject by preparing women for careers as child nutritionists. To gain respect in this sub-specialty, home economists allied themselves with pediatricians, hoping to persuade the medical community to cooperate with their graduates. As such, Deans of home economics faculties asked pediatricians to teach their nutrition students the signs of malnourishment in children and to continue to supervise the graduates' work in nutrition classes across America (Kohene 1922).

Many of these women were influenced by the work of Boston pediatrician Dr. W. P. Emerson who developed height and weight standards to assist physicians in determining the health and nutrition status of children. Children who were ten per cent or more underweight were considered malnourished and Emerson placed heavy emphasis on regular eating and sleeping habits fostered through home discipline, rather than the composition of the diet. When women nutrition workers in Newfoundland attempted to apply these methods, emphasizing the need for maternal efficiency, their efforts met with little success. Once they came to understand the material circumstances of outport women who had enormous responsibilities during the fishing season, the nutrition workers realized that the women had little time for nutrition classes. Where nutrition campaigns were most successful they were tied to the provision of desperately needed relief, which was the case in the White Bay area in 1921. After three poor fishing seasons and the failure of a lumbering business, families were desperately in need of the foodstuffs provided at the nutrition classes. Grenfell workers also promoted gardening as a means of adding nutritious food to the local diet. Viewed as women's work, gardening and canning promised more self-sufficiency, one of Wilfred Grenfell's major concerns about Newfoundland (Lush in progress).

Government buy-in to the desirability of employing a professional nutritionist occurred in the context of the Cuthbertson report and policy discussions around a nutrition campaign. By October 1946 the campaign included a nutrition advisor, Miss Flora Russell, who had been seconded from the Imperial Bureau of Animal Nutrition in the United Kingdom and who had worked with Cuthbertson. She was the first of several full time nutrition advisors hired by government. Russell confirmed the prevalence of deficiency diseases reported by others and she recommended working with existing groups, including Memorial College where she hoped to have nutrition classes to train a team of workers. Remaining in Newfoundland until September 1947, Russell reported on the strides made in implementing policy. In her first report she noted the improvement in the cocoa milk formula for schools, the steps taken to import concentrated orange juice, the experiments with enriched white flour, the launch of a twice-weekly radio program and the planned employment of another nutritional advisor. In her second and final report she listed the increased distribution of milk powder and cod liver oil and the steps taken to provide nutrition education in the schools, to young mothers through public health nurses, and to the general population through adult education workers (Newfoundland Nutrition Council 1945).

Newfoundlander Ella Brett became the nutritional advisor to the Department of Public Health and Welfare following Russell's departure. Trained in home economics, Brett continued the education activities including the radio program "Kitchen Corner" inaugurated in 1947. Brett also prepared columns for the print media. In 1951 she was replaced by Alberta-born Olga Anderson who remained as nutritional advisor until her retirement in 1979. Anderson was not only trained in home economics but she also had a degree in public health. These women, from Edna Baird in the early forties to Olga Anderson through the 1970s, helped shape the educational campaign envisioned by the Newfoundland Medical Association and Dr. Cuthbertson in the 1940s.

As part of the process of "modernizing" Newfoundland women and thus effecting a change in dietary habits, nutritionists used the media to encourage change. The radio program, "Kitchen Corner," ran from 1947 to 1968 when it became a television show. The program carried the message of good nutrition into the homes of women, featuring topics that dealt with particular types of food, their nutritional value, how to cook and preserve them as well as public health issues. Frequently using guest experts to convey these messages the programs also discussed gardening, dental health, the roles of various volunteer groups, childhood diseases, and pregnancy. Program topics were repeated in recognition that food habits were slow to change. Although the nutrition consultant prepared all the shows each script had to be approved by a medical officer in the Department even though doctors were quick to point out their lack of qualifications and time to carry out the work properly (*Kitchen Corner* 1947–68).

Newspaper and magazine columns were also popular venues for nutrition experts who couched their advice as wise consumerism based on science. In March 1945 Edna Baird's column stressed the importance of proper planning: "When the food budget is a problem, time must be spent instead of money because there is no getting around the fact that it takes a good deal of time to plan nutritious meals at low cost. It is easier for homemakers who know food values," she wrote. In Baird's opinion malnutrition resulted not from poverty alone but also from lack of knowledge (Baird 1945). Five years later Brett reiterated the valuable role women played: "The homemaker must know how to plan her family's meals; Nutritionists, Health Educators, the radio and the press all try to convince her that her choice of food for her family will largely determine not only their good health but also their good looks . . . The homemaker's task then would seem to be an important one. The family food is vital to family health and important to happiness." Further on she warned that overcooking could destroy over half the food value of vegetables. Appealing for acceptance of the "scientific view" Brett asked women whether they would believe the studies of scientists or succumb to promises of freedom from drudgery that some advocated through the use of canned and convenience foods. In another article she remarked on the importance of food habits in childhood for later health thus suggesting again the responsibility of mothers for their children's well-being (Brett 1950).

In her career as nutritional advisor Anderson continued the education program as outlined in 1947 but she also expanded the role. Consultative services were provided to a wide range of government, professional, and voluntary groups. For example in 1957 her annual report showed her work with nurses and other health professionals as well as with the 4-H program (clubs for rural children founded in the United States in the early twentieth century), teachers, home economics groups, health inspectors, and midwives. The press and radio absorbed much of her time as Anderson prepared the "Kitchen Corner" scripts, a bulletin called "Newfoundland Nutrition Notes," and numerous pamphlets and articles (Anderson 1957). In other years she also worked with the Newfoundland Tuberculosis Association, the Red Cross, the Jubilee Guilds, the Girl Guides and the Victorian Order of Nurses (Anderson 1956). As a nutritionist Anderson's work also involved surveys and studies, such as the three-day food habits study of 135 expectant mothers from 1963 to 1965 and the research done for Memorial University's hypertension study during 1967–68 (Anderson 1967). School children's food habits were also studied in the 1960s and 70s. During these years government employed more nutritionists, a number of them in other government departments and some of whom were assigned to cover particular geographical areas in the province. By the mid-1970s, with the publication of the nationwide Nutrition Canada Report, Anderson found herself acting on a somewhat revised set of recommendations which emphasized consumption of milk for all age groups (not just children), more iron rich foods, more physical activity, increased protein for preschoolers, teenage girls, pregnant women, and the elderly as well as the use of Vitamin D fortified milk. Nutrition for school children, infants, and pregnant women became a priority and remains so today in government policy (Newfoundland Department of Health, n.d.).

CONCLUSION

As Anderson stepped down and retired in 1979 she no doubt saw significant changes in the diet and health of Newfoundlanders over the course of three decades of work. In the early 1950s when she began her career Newfoundland had recently joined Canada and begun moving along the path of modernization. As salted fish was replaced by fresh frozen products, women no longer played a key role in salting and drying fish but were increasingly portrayed as modern housewives, as consumers with responsibility for the purchase, preparation, and serving of meals. Professional expertise in nutrition played a larger role as government turned to nutritionists to help solve the perceived malnutrition of a "traditional" and impoverished island. By 1968, however, the government no longer distributed free cod liver oil and orange juice except in Labrador. By the mid-70s nutrition had a higher profile thanks to the nationwide study, the creation of an Interdepartmental Committee on Nutrition within the provincial government bureaucracy and the establishment of a nutrition program at Memorial University. Health care providers and nutrition experts no longer faced the problems of dietary deficiency diseases so much as the growing awareness that children were eating unhealthy foods and that cardio-vascular problems threatened many adults who lived on the traditional Newfoundland diet high in salt and fat content (Fodor et al. 1973). Despite the recognition in the 1970s of the importance of nutrition in influencing health, by the late 1980s and into the 1990s nutritionists were removed from active roles in public health and reassigned to other duties, only some of which related to nutrition.

Ironically, nutrition has in the last few years come back into sharp focus not only through debates over obesity among children and discussions of lifestyle changes that improve health but also in renewed concern for food security and food safety in a global context. While we rarely see cases of scurvy and rickets in North American society today, these and other dietary deficiency diseases are more common in other parts of the world where access to food is a chronic problem whether caused by nature or human initiatives. In addition North Americans still face the consequences of chronic diseases that can be related to poor nutrition. Lack of food security—"access to sufficient, safe, nutritious and culturally appropriate food" (World Food Summit 1996)—still faces those who live in poor economic circumstances in less wealthy regions (such as Atlantic Canada) or in rural, isolated communities where prices for food are often higher (such as Labrador or outport Newfoundland). The research presented in this chapter reminds us that past and present are always connected; in both the past and the present nutrition and its relationship to health have been the subject of research and policies that have not always served the needs of those most affected.

ACKNOWLEDGEMENTS

I would like to acknowledge the assistance of Gail Lush, Master's Student at Memorial University in St. John's, Newfoundland.

In: Resetting the Kitchen Table
Editors: C. C. Parrish et al., pp. 191-198

ISBN 1-60021-236-0
© 2008 Nova Science Publishers, Inc.

Chapter 13

CONCLUSIONS: WHAT FOOD SECURITY IN COASTAL COMMUNITIES REALLY MEANS

*Christopher C. Parrish[1], Nancy J. Turner[2], Rosemary E. Ommer[3]
and Shirley M. Solberg[4]*

[1]Ocean Sciences Centre, Memorial University of Newfoundland
[2]Environmental Studies, University of Victoria
[3]University SSHRC Grants Facilitator and Adjunct Professor, History,
University of Victoria
[4]Nursing and Women's Studies, Memorial University of Newfoundland

ABSTRACT

For most people in the modern world their food is found on the shelves of a supermarket. It comes sanitized in a packaged form, with additives and preservatives, making it safe and personally acceptable, but how nutritious and culturally appropriate is it? Has it been produced in ways that are environmentally sustainable and that protect domestic food production or food sovereignty? In this concluding chapter we begin with the history of human food production and then we examine the components of the World Food Summit's definition of food security for the Atlantic and Pacific coasts of Canada as well as the wider implications. We look at what food security in coastal communities really means in terms of policy, nutrition, health, and well-being for people in these communities leading us to the following revision of the food security definition:

Food security exists when all people, at all times, have access to enough nutritious, safe, personally acceptable, and culturally appropriate foods, produced in ways that are sustainable and that protect domestic food production.

INTRODUCTION

Food is a fundamental human need, and food security is recognized as a critical issue in communities all over the world. In this volume we have examined food security as it applies to communities on Canada's east and west coasts. We have found that we can use the World Food Summit's definition of food security (Turner et al., Chapter 1, this volume) to see when food security exists; however, here we explore the means of getting there and what the social-ecological implications of food security may be. This book examines many aspects of food security in coastal communities: historical, cultural, nutritional, and educational. Most importantly, it connects food security to changing economies and changing ecosystems. "Resetting the kitchen table" is a metaphor for changes that are occurring in families and communities not only in daily food consumption but also in local availability of food, and hence in local culture as expressed in local practices for producing, obtaining, processing, and serving foodstuffs. Although we use Canadian examples in order to identify the pathways between regional and national systems, our findings are relevant to all resource-based coastal communities in the developed world, and probably also beyond that to rural communities everywhere. We have examined some of the basic nutritional requirements for human health and well-being, and the important links between and among environment, lifestyles, economies, social structures, cultures and institutions, knowledge acquisition and transmission, biodiversity conservation, and human nutrition.

HISTORICAL PATHWAYS

Prior to contact with Europeans, aboriginal peoples on the Pacific Northwest coast of North America held practices and belief systems conducive to long-term conservation of wild resources, such as establishing proprietorship and allowing long-term monitoring of resource populations, managing habitats to create more food resources, selective harvesting by size and life stage to maintain the reproductive capacity of plant and animal populations, and social controls against wasting food or harvesting more than was needed (Haggan et al., Chapter 4, this volume; Anderson 2005; Deur and Turner 2005; Turner 2005). Although there is a paucity of other quantitative evidence for long-range conservation by humans, culture has been a way of embedding successful practices in society, and our work has demonstrated such characteristic features of cultural heritage in coastal communities on both coasts (Ommer and team, in press).

In the rich coastal environment of both coasts, nutrition was a built-in part of the seasonal round of hunting, fishing, and gathering wild berries. The more coastal communities have become drawn into the 'modern' world, the less subsistence plays a part and the more the dietary habits of urban society have penetrated their existence (Ommer 2000). In the days of a subsistence economy, food security had been assured provided there was a very small cash input with which to purchase the means of production (shovel, seed, fishing line). Even in the great Depression of the 1930s, when wage jobs were hard to come by, people moved back from the cities of mainland Canada to the outports of Newfoundland, explaining that "you'll never starve here." That's a graphic description of the advantage of a subsistence over a waged economy way of life.

With the establishment of a modern capitalist economy came wage labour and a dependence on cash. In other words, food production and food consumption were severed and the quality of person's diet was thereafter tied to income, as it is in urban societies. Today, if you have to buy food and you do not have an income, you would starve without protection from state social welfare provisions. Food is less, not more, secure now than it was in the 1930s and its quality is often open to question: donuts, fried chicken, and sugared drinks.

On the West Coast, First Nations communities worked hard for their food, but in general, theirs was a nutritious and diverse diet, with plenty of seafood, game, and seasonally available greens, roots, and berries which provided them with all of the essential vitamins and minerals, as well as dietary fibre. The elders of today remember their traditional foods with great pleasure, and enjoy whatever they can get of these foods. It has been increasingly difficult for them to acquire traditional foods, however. Many people still long for the time when they spent weeks and months out on the land, gathering and preparing their own food, as the most important activity. The foods are still there, at least some of them, but the way of life that knows how to get them and process them is slipping away. This is clearly voiced in Helen Clifton's quotation with which we started this book (Turner et al., Chapter 1).

PHYSICAL AND ECONOMIC ACCESS TO FOOD

Today a planet-wide 'nutrition transition' exists in which many different human populations are experiencing major shifts away from locally produced, culturally distinctive, nutritious foods to marketed, highly processed foods of lower nutrient density. While this is true in both the so-called 'developed' and the less-developed world, not enough scholarly attention has been paid to the small communities of the First World, which are often suffering the kind of distress that makes them akin to their less-developed world counterparts. They, too, are faced with the circumstances of the nutrition transition, where most of the food they consume now comes from supermarket shelves, albeit through phone orders and long distance transport to remote places. Canada's east and west coastal communities provide particularly telling case studies because the collapse of the ground fishery in the NW Atlantic, and a serious decline in salmon and other marine food species in the Pacific Northwest, coupled with major economic and social change, have generated a major crisis in environmental and social well-being.

Food production and availability are dependent on many factors, both social and environmental, and in many communities throughout the world, social and economic restructuring has compromised the ability of people to meet their nutritional needs. Access to food can be very different within a region and even within a community, with the elderly and poor being particularly vulnerable (Solberg et al., Chapter 11, this volume). Within the study area, we have indeed documented a shift away from harvesting and consumption of local food and towards the commodification of food resources in a global economy, with local communities depending more and more on marketed food produced outside their region. Ironically, this food is often purchased with wages from commercial food harvest or processing within their region.

Our dependence on food produced outside our regions brings with it an economic vulnerability to world prices for many things, such as fossil fuel, and other uncertainties associated with global transportation systems. The production and distribution of crops, grazing animals, and fishing in developed countries are currently highly dependent on energy in the form of oil as compared to pre-industrial times when some of these were very much local activities, although the fishery on the East Coast has been based on long-distance international trade for centuries (Innis 1954; Ommer 1991, 141–175). Gunther (2001) provides strong arguments for moving back towards the earlier situation of more local production in order to increase food security, by simply reducing energy demands. He proposes that food production be integrated with settlements ('ruralization') so that most of the food needed by a population is produced close at hand, and nutrients are recycled within that region. Others, such as those in the Slow Food movement, have argued for increased 'bioregionalism' in food production in order to maintain the diversity and local character of peoples' food systems (Petrini 2005). There is clearly much to be said for such a strategy, but it will be difficult to achieve in a world where the drive towards urbanization is global and rapid, and governments think of it as improving the efficiency of the state. The evidence provided by this book, and other writing, is in stark contrast to such thinking.

FOOD SAFETY AND NUTRITION

Foods are consumed for their distinct flavour and taste as well as for their nutritional and potential health benefits. However food safety can be compromised by contamination with microorganisms or toxic compounds. In this way the safety and nutritional value of a foodstuff can be completely separate—low quality foods can still be safe to consume—or intimately interconnected as in the case of fats and fat-soluble contaminants. Both the type and amount of lipid in a food are important in determining its nutritional value as well as the potential accumulation of persistent organic pollutants (POPs) such as polychlorinated biphenyls (PCBs), which remain in the environment despite restrictions on their use. Such compounds are distributed globally and are often found in household products such as butter, with the highest concentrations occurring in Europe and North America (Kalantzi et al. 2001). PCBs even occur in ooligan grease prepared by First Nations families in British Columbia, albeit at concentrations that are below the regulation limits established by Health Canada (Chan et al. 1996). While use of some POPs is restricted through regulation, others, such as polybrominated diphenyl ethers (PDBEs), are widely used in a variety of products as flame retardants and also have a global distribution in tuna (Ueno et al. 2004). People who consume fish are thus presented with a dilemma, having to balance the positive effects of omega-3 fatty acids, as described in Chapter 1, with the amount of POPs dissolved in the fish lipids (e.g. Jacobs et al. 2004). Unfortunately this dilemma is not limited to POPs; mercury is also a contaminant of finfish and shellfish, although several species that have the highest content of omega-3 fatty acids have low mercury concentrations (Mahaffey 2004). The sub-polar oceanographic climate zone of Newfoundland and Labrador ensures high levels of omega-3 fatty acids in local seafood (Parrish et al., Chapter 3, this volume), which is still an important component of people's diet despite changes in the fisheries (Solberg et al., Chapter 11, this volume). The contaminants in the fats and oils of fish are just one example of the complexities facing us in characterizing, determining, and trying to improve human food security.

FOOD PREFERENCES AND CULTURALLY APPROPRIATE FOOD

No matter how good a given food is nutritionally, it offers no sustenance if it is not eaten. In places where food choice is abundant, palatability becomes a major factor in determining which food is eaten and therefore which nutrients are ultimately consumed. The palatability of a food includes the texture or 'mouthfeel' and secondly the flavour, which is determined by the taste in the mouth and aroma in the nose. However, what is acceptable taste-wise and sufficiently agreeable in flavour to be eaten also depends on cultural traditions and familiarity, usually cultivated from early childhood. Therefore, understanding cultural constraints on food, and hence nutrition, is an important area of health research. Likewise, deprivation of culturally important foodstuffs can lead to loss of culturally identity, stress, and loss of appetite, especially among older people. By contrast, culturally appropriate food, food preferences, and good nutrition are often positively correlated. For example, residents of the Great Northern Peninsula in Newfoundland prefer the flavour of moose over beef (Solberg et al., Chapter 11, this volume). Moose is readily available in the area and is a low fat source of protein. In addition, moose hunting is good for social cohesion and encourages sharing in the community as part of the informal economy (Ommer et al. Chapter 8, this volume).

A major advantage of global food circulation has been that people are exposed to a wider range of culinary specialties than before, but that is negated when the global marketplace deals in large quantities of fast foods and a dearth of fresh locally produced food. In China and elsewhere, including some Canadian communities, obesity is now becoming a problem as the fast food corporate giants penetrate global markets.

LINKING FOOD SECURITY AND EDUCATION

From school age through college and university, we teach the importance of good nutrition and healthy food choices, yet our students are surrounded by vending machines displaying foods and beverages of minimal nutritional value, high in calories and fat. Ongoing funding challenges in public education push this dichotomy further through the use of sponsored education materials that at their extreme can amount to nutrition lessons taught by fast food and multinational beverage companies. Similarly, and equally disturbing, are widely distributed educational resources in which the discussions of environmental issues are carried out by oil companies (Rockne 2002).

On the positive side, school provision of nutritious meals and snacks to needy students improves academic achievement, social behaviour, and physical and emotional well-being, as found in northern Vancouver Island (Harris and Shepherd, Chapter 9, this volume). In addition, First Nations teachers and support workers in the inner-city community school included in the study try to involve the children in special food celebrations of their own cultures and this enhances their cultural identity and self esteem.

Education in nutritional values has been very important in Newfoundland and Labrador (Kealey, Chapter 12, this volume), and in British Columbia, school food programs run by First Nations teachers provide one way for aboriginal youth to learn traditional ways of food gathering and preparation (Tirone et al. Chapter 10, this volume). The next step in food security education should clearly also involve an environmental systems approach (e.g. Watt

1982) where conservation, and environmental stewardship are emphasized along with good nutrition, and social responsibility (Petrini 2005).

LINKING HUMAN FOOD SECURITY AND FOOD WEBS

Human food gathering generally obeys the rules of optimal foraging theory within the constraints of existing foraging technology. We now have the technical ability to catch whatever species is abundant within an ecosystem, eliminating refuges enjoyed by earlier animal populations. Thus, pre-industrial fisheries were sustainable mainly because of our inability to access a major part of the exploited stocks (Pauly et al. 2005). Notably, however, some indigenous societies had the capability to decimate salmon stocks through establishing weirs, and fishtraps on virtually all spawning streams, but chose not to do so (Anderson 1996). Conversely, there were already signs of over-exploitation of the centuries old fishery off Newfoundland at the beginning of the eighteenth century, which prompted the expansion of the offshore banks fishery (Martinéz Murillo and Haedrich, Chapter 2, this volume). Our invasion of the refuges previously provided by distance, and depth has led to a decline in global fish landings. Other changes in food availability may relate to shifts in oceanographic conditions (Kennedy, Chapter 6, this volume) causing some species to become rare, and others to acquire new importance. Meanwhile, conservation legislation has clearly been critical to wildlife survival (Montevecchi et al., Chapter 7, this volume), and key habitat preservation (e.g. Duarte 2002).

Declines in fish landings may be exacerbated by farming carnivorous fish such as salmon (Volpe Chapter 5, this volume). However, it is important to distinguish industrial fisheries capturing 'feed fish' such as sandeel, and sprat for use mainly in aquaculture, and livestock rearing from those producing 'food fish' for direct human consumption. The worldwide feed grade fisheries are now considered to be generally quite well managed although there are clearly areas for improvement (Shepherd et al. 2005). Nonetheless, there is a strong need for the re-establishment of refuges for all marine species (Pauly et al. 2005), and the development of fishmeal production from seafood waste or discards (Shepherd et al. 2005). This, and the continued increase in the proportion of plant ingredients in fish diets, will mitigate pressure on feed fish stocks.

Refuges can take the form of marine protected areas; however, they need to protect not only nursery grounds but migration routes as well (Haggan et al. Chapter 4, this volume). Furthermore, fishing gear needs to be modified to minimize by-catch, and physical damage to other species or their habitats. Habitat protection, as well as protection through no-catch zones, must be a primary focus. Eelgrass (*Zostera marina*) beds, for example, which are the 'nurseries' for myriad fish and other marine organisms important in the food web, have suffered marked restriction and deterioration in the past few decades in many areas, due to pollutants, eutrophication, shading, and various kinds of disturbance (*cf.* Thom et al. 1998; Wyllie-Echeverria and Ackerman 2003). Other tidal marsh rhizomes and bulbs have traditionally been cultivated in the Pacific Northwest. Large gardens of highly productive estuarine roots were very important sources of carbohydrates and other nutrients for the indigenous peoples (Deur 2002; Deur and Turner 2005).

POLICY AND FOOD SECURITY FOR COASTAL COMMUNITIES

In this book we have demonstrated many of the intricate and complex interrelationships between environmental change, economic restructuring, ecosystem health, and human health as these relate to food production, consumption, and availability. Our work is intended to provide direction for reassessing policy and regulation in First World countries with a view to developing greater environmental and social sustainability for coastal communities, through increased food security and hence resilience.

Aquaculture is now the fastest growing food-producing industry in the world and is drawing criticisms from environmental groups and consumers on a number of issues. While those who support it see it as a potential cornerstone for the provision of nutritious and culturally appropriate food to coastal communities, others are concerned that it is unwise to promote those aquaculture operations where inadequate research has been done to demonstrate their benign environmental impact or to establish ways to reduce their present environmental costs significantly. Volpe (Chapter 5, this volume) has focused on some major environmental concerns related to farming with Atlantic salmon along the coastal waters of British Columbia. The repercussions and impacts on wild salmon, including the biological threat of sea lice epidemics, are proving to be more deleterious to Pacific salmon than was ever anticipated. Lessons can be learned, too, from one hundred years of steel making in Sydney, Nova Scotia, which played a pivotal role in the economy, history, and culture of the surrounding communities. The government-owned corporation was closed in 2001, and now three levels of government are working on a contaminated estuary that may require the largest, most expensive, toxic cleanup in Canadian history. By the same token, large scale high technology vessels that drag the bottoms of the oceans and destroy marine habitats are efficient only in the short term. Ecosystems function in a complex of interrelated checks and balances and destroying key parts of that lattice will have, over the longer run, serious consequences for marine foodstuffs.

Today we have a range of integrated coastal management initiatives in operation in Canada. We discuss these at length elsewhere (see Chapter 15 of Ommer and team, in press) in terms of their ability to maintain, restore, or improve coastal ecosystems and their human societies. This is a management process for negotiating and implementing public policy to achieve sustainable coastal development (Olsen 2003), but co-management, and a greater respect for and recognition of local knowledge and interests will be required if stewardship is really to be achieved.

CONCLUSION

In this book we have sought to provide an unusual combination of case studies drawn from multiple disciplines whose analyses combine to demonstrate the interconnectedness of food systems, and of social and ecological factors influencing food and nutrition in communities on both the Atlantic and Pacific coasts of Canada. We have looked at the biophysical aspects of coastal foods and have connected these to human food security. We have dealt with food webs and fish-fisheries dynamics, nearshore marine food webs and human impacts, food web models and requirements for restoring marine ecosystems to some level of their past abundance, and aquaculture practices and their effects on ecological integrity and food security.

We have also addressed some of the key dynamics of livelihoods and food production as they relate to coastal communities and their history. We have provided case examples of past and present hunting practices on the Atlantic coast and relationships to environmental and social restructuring. We have looked at informal economies and subsistence strategies that have helped to sustain coastal communities, and their contributions to resilience. We have also examined the history of food security and its relationship with restructuring.

Finally we have covered the cultural aspects of food production, food use and food security. We discuss the relationship between public school education, nutrition, and health, and link food with cultural identity and social health, providing case examples of the ongoing importance of food in feasts, ceremonies, and celebrations, and emphasizing the relationships between cultural health and social health and economic restructuring.

The book was written for any academic researcher or policy maker working on any aspect of food security in coastal communities. However, we hope that in the end, all those with an interest in food and the current global nutrition transition, will find this book useful and helpful in understanding the complexities of food production and consumption and its cultural and environmental components. We opened this chapter with a definition of food security that is based on the World Food Summit definition (Chapter 1) but which adds critical issues associated with production. Our work underlines the importance of food production being environmentally sound and socially just.

We cannot go back to the old ways, nor should we. But, there are lessons that we can learn from those days. Sustenance based on local food production would require communities becoming embedded in their environments again. When we consider problems of ecosystem collapse and terrestrial degradation, it might well pay us to ensure that some of us regain that intimate knowledge of local environments. Modern technology can be used to make us the most efficient killers on the planet, or it could be used to turn us into a society that uses its skills to sustain itself and the environment that supports it. Whether we think food comes from a supermarket shelf or not, the truth is that ultimately, it comes from natural systems which we are steadily damaging. We need a society in which some of us are stewards of the environment, and our coastal communities, with their great desire to remain in the places that they have called home for generations, are prime candidates for the task. This is not a matter of rural romantic yearning for past times, but a forward looking vision that would give us vibrant, technologically sophisticated, and diversified communities in tune with ecosystem values which we are perilously close to losing. Tying stewardship and food security to the value of ecosystem services is a way for coastal communities to produce a model for the modern world.

REFERENCES

Ackman, R. G. (2003). A history of fats and oils in Canada. *Lipids,* 38, 299–302.

Ackman, R. G. (1989). Nutritional composition of fats in seafoods. *Progress in Food and Nutrition Science,* 13, 161–241.

Adamson, J. D., Joliffe, N., Kruse, H. D., Lowry, O. H., Moore, P. E., Platt, B. S., Sebrell, W H., Tice, J. W., Tisdall, F. F., Wilder, R. M., & Zamecnik, P. C. (1945). Medical survey of nutrition in Newfoundland. *Canadian Medical Association Journal,* 52, 227–250.

Agular, A. 1986. A review of old Basque whaling and its effect on the right whales of the North Atlantic. *Reports of the International Whaling Commission,* 10, 191–199.

Ainsworth, C., Heymans, J. J., & Pitcher, T. J. (2004). Policy search methods for Back to the Future. In T. J. Pitcher (Ed.), Back to the Future: Advances in methodology for modelling and evaluating past ecosystems as future policy goals. *Fisheries Centre Research Reports,* 12(1), 48–63.

Ainsworth, C., Heymans, J. J., & Pitcher, T. J. (2002). Ecosystem models of Northern British Columbia for the time periods 2000, 1950, 1900 and 1750. *Fisheries Centre Research Reports* 10(4), 41pp.

Ainsworth, C., & Pitcher, T. J. (In press.). Back-to-the-Future in Northern British Columbia: Evaluating historic marine ecosystems and optimal restorable biomass as restoration goals for the future. *Proceedings of the Fourth World Fisheries Congress.* May 2–6, 2004. Vancouver, Canada.

Ainsworth, C., & Pitcher, T. J. (2005a). Using local ecological knowledge in ecosystem models. In G. H. Kruse, V. F. Gallucci, D. E. Hay, R. I. Perry, R. M. Peterman, T. C. Shirley, P. D. Spencer, B. Wilson & D. Woodby (Eds.), *Fisheries assessment and management in data-limited situations.* University of Alaska Fairbanks: Alaska Sea Grant College Program.

Ainsworth, C., & Pitcher, T. J. (2005b). Evaluating marine ecosystem restoration goals for Northern British Columbia. *Assessment and Management of New and Developed Fisheries in Data-Limited Situations: Proceedings from the 21st Lowell Wakefield Symposium.* Alaska Sea Grant.

Ainsworth, C., & Sumaila, U. R. (2005). Intergenerational valuation of fisheries resources can justify long-term conservation: a case study in Atlantic cod (*Gadus morhua*). *Canadian Journal of Fisheries and Aquatic Sciences,* 62, 1104–1110.

Akenhead, S. A., Carscadden, J., Lear, H., Lilly, G. R., & Wells, R. (1982). Cod–capelin interactions off Northeast Newfoundland and Labrador. In M. C. Mercer (Ed.),

Multispecies Approaches to Fisheries Management Advice. Canadian Special Publication, *Fisheries Aquatic Sciences,* 59, 141–148.

Alphonso, C. (2004, October 21). Ontario bans junk food from school machines. *Globe and Mail*, p. A9.

AMEC, Earth and Environmental. (2002). Aquaculture information review—an evaluation of known effects and mitigations on fish and fish habitat in Newfoundland and Labrador. *Canadian Technical Report on Fisheries and Aquatic Sciences,* 2434, vii.

Amilien. V. (2001). *What do we mean by traditional food? A concept approach.* Working paper No.1-2001. National Institute for Consumer Research, Norway.

Anderson, E. N. (1996). *Ecologies of the heart. Emotion, belief and the environment.* New York: Oxford University Press.

Anderson, M. K. (2005). *Tending the wild. Native American knowledge and the management of California's natural resources.* Berkely, CA: University of California Press.

Anderson, M. S. (Ed.). (1984). *The Tsimshian. Images of the past, views for the present.* Vancouver, BC: UBC Press.

Anderson, M. S. (2004). Understanding Tsimshian Potlatch. In R. B. Morrison & C. Roderick Wilson (Eds.), *Native peoples: the Canadian experience* (3rd ed., Chapter 24). Toronto, ON: Oxford University Press.

Anderson, O. (1967). Three day food habits study of 135 expectant mothers. *Canadian Nutrition Notes* 23, 7, 73–75.

Anderson, O. (1957). Nutrition activities in Newfoundland—1957. *Nutrition Notes.* 14, 8, 57–60.

Anderson, O. (1956). Nutrition services in Newfoundland. *Nutrition Notes,* 12, 6, 41–42.

Anderson, G. L., & Jones, F. (2000). Knowledge generation in educational administration from the inside out: The promise and perils of site-based, administrator research. *Educational Administration Quarterly,* 36, 428–464.

Appleton, V. B. (1921). Observations on deficiency disease in Labrador. *American Journal of Public Health,* 11, 617–621.

Ardron, J. A. (2005). *Protecting British Columbia's corals and sponges from bottom Trawling.* Sointula, BC: Living Oceans Society *http://www.livingoceans.org/fisheries/research.shtml.*

Ardron, J. A. (2002). A recipe for determining benthic complexity: An indicator of species richness. In J. Breman (Ed.), *Marine Geography: GIS for the Oceans and Seas* (Chapter 23). Redlands, CA: ESRI Press.

Atasoy, Y., & Carroll, W. K. (Eds.) (2003). *Global shaping and its alternatives.* Aurora, ON: Garamond Press.

Aure, J., & Stigebrandt, A. (1990). Quantitative estimates of the eutrophication effects of fish farming on fjords. *Aquaculture* 90, 135–156.

Aykroyd, W. R. (1930). Beriberi and other food-deficiency diseases in Newfoundland and Labrador. *Journal of Hygiene*, 30, 357–386.

Aykroyd, W. R. (1928). Vitamin A deficiency in Newfoundland. *Irish Journal of Medical Science,* 28, 161– 165.

Ayres, R. V., van der Bergh, J. C., & Gowdy, J. M. (2001). Strong versus weak sustainability: Economics, natural science and "consilience." *Environmental Ethics,* 23, 155–168.

Baird, E. (1945). Food and your health. *Newfoundland Government Bulletin,* 19 March.

Bannister, K. (2003). *Community-university connections: Building a foundation for research collaboration in British Columbia*. Working Paper Series, Clayoquot Alliance for Research, Education and Training. University of Victoria.

Barnabé, G. (1994). Fish nutrition. In G. Barnabé (Ed.), *Aquaculture: Biology and Ecology of Cultured Species* (pp. 246–270). New York, Toronto: E. Horwood.

Barrett, G. W. (1985). A problem-solving approach to resource management. *Bioscience, 35*, 423–427.

Baum, J. K., Myers, R. A., Kehler, D. G., Worm, B., Harley, S. J., & Doherty, P. A. (2002). Collapse and conservation of shark populations in the Northwest Atlantic. *Science, 299*, 389–392.

Bax, N. J. (1991). A comparison of the fish biomass flow to fish, fisheries, and mammals in six marine ecosystems. *ICES Marine Science Symposia, 193*, 217–224.

Bax, N. J. (1998). The significance and prediction of predation in marine fisheries. *ICES Journal of Marine Sciences, 55*, 997–1030.

BC Ministry of Education (2006). *School Performance Report 2000/01 to 2004/05. Port Hardy Secondary, School Code 08585026, Vancouver Island North*. Retrieved February 9, 2006 from: http://www.bced.gov.bc.ca/reports/pdfs/student_stats/08585026.pdf.

BC Ministry of Education (2005). *Student Statistics 2001/02 – 2005/06. Port Hardy Secondary, School Code 08585026, Vancouver Island North*. Retrieved February 9, 2006 from: *http://www.bced.gov.bc.ca/reports/pdfs/student_stats/08585026.pdf.*

Beaudry, M., Hamelin, A-M., & Delisle, H. (2004). Public nutrition: An emerging paradigm. *Canadian Journal of Public Health, 95*, 375–377.

Beddington, J. R. (1984). The response of multispecies systems to perturbations. In R. M. May (Ed.) *Exploitation of Marine Communities* (pp. 209–225). Berlin: Springer-Verlag.

Beddington, J. R., Arntz, W. E., Bailey, R. S., Brewer, G. D., Glantz, M. H., Laurec, A. J. Y., May, R. M., Nellen, W. P., Smetacek, V. S., Thurow, F. M. R., Troadec, J-P., & Walters, C. J. (1984). Management under uncertainty. In R. M. May (Ed.), *Exploitation of Marine Communities* (pp. 227–244). Berlin: Springer-Verlag.

Bella, L. (1992). *The Christmas imperative: Leisure, family and women's work*. Halifax, NS: Fernwood Publishing.

Belloc, H. (1896). *The bad child's book of beasts*. Alden: Oxford.

Ben-Dor, S. (1966). Makkovik: Eskimos and settlers in a Labrador community. St. John's: ISER, *Newfoundland Social and Economic Studies*, No. 4.

Berkeley, S. A., Chapman, C., & Sogard, S. M. (2004). Maternal age as a determinant of larval growth and survival in a marine fish, *Sebastes melanops. Ecology, 85*, 1258–1264.

Berkes, F. (1999). *Sacred ecology: Traditional ecological knowledge and resource management*. Philadelphia, and London: Taylor and Francis.

Berkes, F. T., & Turner, N. J. (2006). Knowledge, learning and the resilience of social-ecological systems. *Human Ecology, 34*(4) in press

Beverton, R. J. H., Cooke, J. G., Csirke, J. B., Doyle, R. W., Hempel, G., Holt, S. J., MacCall, A. D., Policansky, D. J., Roughgarden, J., Shepherd, J. G., Sissenwine, M. P., & Wiebe, P. H. (1984). Dynamics of single species. In R. M. May (Ed.), *Exploitation of Marine Communities* (pp. 13–58). Berlin: Springer-Verlag.

Bianchi, G. H. Gislason, K. Graham, L. Hill, X. Jin, K. Koranteng, S. *et al.* 2000. Impact of fishing on size composition and diversity of demersal fish communities. *ICES Journal of Marine Science, 57*, 558–571.

Birch, L. L. (1999) Development of food preferences. *Annual Review of Nutrition*, 19, 41–62.

Birkeland, C., & Dayton. P. K. (2005). The importance in fishery management of leaving the big ones. *Trends in Ecology and Evolution,* 20, 356–358.

Bjorklund, H., Bondestam, J., & Bylund, G. (1989). Residues of oxytetracycline in wild fish and sediments from fish farms. *Aquaculture,* 86, 359–367.

Bjorkstedt, E. P. (2000). Stock-recruitment relationships for life cycles that exhibit concurrent density dependence. *Canadian Journal of Fisheries and Aquatic Science,* 57, 459–467.

Blackmore, J. (1996). Doing 'emotional labour' in the education marketplace: Stories from the field of women in management. *Discourse. Studies in the Cultural Politics of Education,* 17, 337–349.

Blumenthal, M., Goldberg, A., & Brinckmann, J. (Eds.). (2000). *Herbal Medicine.* Expanded Commission E Monographs. Newton, MA: Integrative Medicine Communications; and Austin TX: American Botanical Council.

Boas, F. (1916) Tsimshian Mythology. *In*: 31st Annual report of the Bureau of American Anthropology for the years 1909-1910. Government Printing Office, Washington.

Bond, S. (2005). Message from the Minister and Accountability Statement, 2005–2008 Service Plan Update, BC Ministry of Education. http://www.bcbudget.gov.bc.ca/sp/educ/.

Bonner, J. T. (1965). *Size and cycle.* Princeton, NJ: Princeton University Press.

Booth, D. (1998). Waist not, want not. In S. Griffiths & J. Wallace (Eds.), *Consuming patterns: Food in the age of anxiety,* (pp. 96–103). Manchester, UK: Mandolin.

Bordajandi, L. R., Gomez, G., Abad, E., Rivera, J., Fernandez-Baston, M. M., Blasco, J., & Gonzalez, M. J. (2004). Survey of persistent organochlorine contaminants (PCBs, PCDD/Fs, and PAHs), heavy metals (Cu, Cd, Zn, Pb, and Hg), and arsenic in food samples from Huelva (Spain): Levels and health implications. *Journal of Agricultural and Food Chemistry,* 52, 992–1001.

Botsford, L. W., Castilla, J. C., & Peterson, C. H. (1997). The management of fisheries and marine ecosystems. *Science,* 277, 509–515.

Bowers, C. A. (1988). *The cultural dimensions of educational computing: Understanding the non-neutrality of technology.* New York, NY: Teachers College Press.

Brashares, J. S., Arcese, P., Sam, M. K., Coppolillo, P. B., Sinclair, A. R. E., & Balmford, A. (2004). Bushmeat hunting and fish declines. *Science*, 306, 1180–1833.

Brett, E., (1950). "Good Food for Good Health" Column. *Atlantic Guardian.* 7, 5 (May), 54–5 and 7,8 (August), 61.

Brice-Bennett, C. (1994). *The dispossessed: The eviction of Inuit from Hebron, Labrador.* Unpublished report submitted to North Program, Royal Commission on Aboriginal Peoples.

Britton, J. (1998). *An evaluation of public involvement in reclamation decision making at three metal mines in British Columbia.* MSc Thesis, Community and Regional Planning, The University of Brtitish Columbia.

Brody, H. (2000). *The other side of Eden. Hunters, farmers and the shaping of the world.* Vancouver, BC: Douglas and McIntyre.

Brody, H., (writer and director). (1994). *The washing of tears.* Nootka Sound and Picture Co. National Film Board of Canada ON#: 9194090.

Bromley, H., & Apple, M. W. (1998). *Education/technology/power: Educational computing as a social practice.* New York, NY: SUNY.

Brookfield, H. C. (1972). *Colonialism, development and dependence. The case of the Melanesian Islands in the South Pacific.* Cambridge, UK: Cambridge University Press.

Brown, L. R. (2000). *Fish farming may soon overtake cattle ranching as a food source.* Aquaculture Network Information Centre. http://aquanic.org/news/2000/farming.htm.

Brown, R. G. B. (1986). *Revised Atlas of Eastern Canadian Seabirds.* I. Shipboard Surveys. Ottawa, ON: Canadian Government Publishing Centre.

Brox, O., (1972). *Newfoundland fishermen in the age of industry.* St. John's, NL: ISER Books.

Brubaker, R., (1984). *The limits of rationality: An essay on the social and moral thought of Max Weber.* London: George Allen and Unwin.

Brunk, C., & Dunham, S. (2000). Ecosystem justice in the Canadian fisheries. In H. Coward, R. E. Ommer, & T. J. Pitcher (Eds.), *Just fish: Ethics and Canadian marine fisheries* (pp. 9–33). St. John's, NL: ISER Books.

Buckworth, R. C. (1998). World fisheries are in crisis? We must respond! In T. J. Pitcher, P. J. B. Hart, & D. Pauly (Eds.), *Reinventing fisheries management* (pp. 3–18). Boston, MA: Kluwer Academic Publishers.

Budge, S. M., Parrish, C. C., Thompson, R. J., & McKenzie, C. H. (2000). Fatty acids in plankton in relation to bivalve dietary requirements. In F. Shahidi (Ed.), *Seafood in Health and Nutrition* (pp. 495–520). St. John's, NL: ScienceTech Publishing.

Bundy, A., Lilly, G., & Shelton, P. (2000). A mass balance model of the Newfoundland-Labrador shelf. *Canadian Technical Report on Fisheries and Aquatic Sciences*, 2310, 117pp + App.

Burke, C., Davoren G. K., Montevecchi, W. A., & Stenhouse, I. J. (2002). *What the past can tell us about the future: Part I Historic reconstruction of coastal settlements and interactions with the marine ecosystem of the Newfoundland Shelf, 1500-2000.* Poster Presentation Ocean Management Research Network (OMRN) Conference, Ottawa.

Burke, C. M., Davoren G. K., Montevecchi W. A. and Wiese, F. K. (2005). Surveys of seabirds along support vessel transects and at oil platforms on the Grand Banks. In P. J. Cransford and K. Lee (Eds.), *Offshore Oil and Gas Environmental Effects Monitoring* (pp. 587-614). Columbus, Ohio: Battele Press.

Byron, G. G. (1812–18). *Childe Harold's Pilgrimage.* Reissued 2004 by Kessinger, Whitefish Montana, 404p.

Caddy, J. F., & Rodhouse, P. G. (1998). Cephalopod and groundfish landings: evidence for ecological change in global fisheries. *Reviews in Fish Biology and Fisheries* 8, 431–444.

Cadigan, S. (1999). The moral economy of the commons: Ecology and equity in the Newfoundland Cod Fishery, 1815–1855. *Labour/Le Travail*, 43, 11.

Calder, W. A. (1984). *Size function, and life history.* Cambridge; MA: Harvard University Press.

Callahan, R. E. (1962). *Education and the cult of efficiency.* Chicago, IL: University of Chicago Press.

Campbell, K. (1984). Hartley Bay, British Columbia: A History. In M. S. Anderson (Ed.), *The Tsimshian. Images of the past, views for the present* (Chapter 1). Vancouver, BC: UBC Press.

Canada. (2004) Canada Gazette Part I, Ottawa, Oct 23.

Carlton, J. T., Geller, J. B., Reaka-Kudla, M. L., & Norse, E.A. (1999). Historical extinctions in the sea. *Annual Review of Ecology and Systematics*, 30, 515–538.

Carpenter, S .R., Chisholm, S. W., Krebs, C. J., Schindler, D. W., & Wright, R.F. (1995). Ecosystem experiments. *Science,* 26, 324–327.

Cartwright, G. (1792). *A Journal of transaction and events during a residence of nearly sixteen years on the coast of Labrador.* Newark, England: Allen and Ridge.

Chaffey, H. 2003. *Integrating scientific knowledge and local ecological knowledge (LEK) about Common Eiders* (Somataria mollissima) *in Southern Labrador.* M.Sc. Thesis, Memorial University of Newfoundland, St. John's, Newfoundland.

Chamberlain, K. (2004). Food and health: Expanding the agenda for health psychology. *Journal of Health Psychology,* 9, 467–481.

Chan, H. M., Berti, P. R., Receveur, O., & Kuhnlein, H. V. (1997). *Dietary intake of arsenic, cadmium, lead and mercury among Dene/Métis in the western Northwest Territories, Canada.* Montreal, QC: 16th International Congress of Nutrition.

Chan H. M., Khoury, M. E., Sedgemore, M., Sedgemore, S., & Kuhnlein, H. V. (1996). Organochlorine pesticides and polychlorinated biphenyl congeners in ooligan grease: a traditional food fat of British Columbia First Nations. *Journal of Food Composition and Analysis,* 9, 32–42.

Chardine, J. W. (2001). Seabird harvest regimes in Canada. In L. Denlinger & K. Wohl (Eds.), *Seabird Harvest Regimes in the Circumpolar Nations (pp. 19–25).* CAFF Technical Report 9.

Chardine, J. W., Collins, B. T., Elliot, R. D., Levesque, H., & Ryan, P. C. (1999). *Trends in the annual harvest of murres in Newfoundland and Labrador.* Bird Trends No. 7, Canadian Wildlife Service Occasional Paper, Ottawa, pp. (11 –14).

Cheung, W. W. L., & Pitcher, T. J. (2004). An index expressing risk of local extinction for use with dynamic ecosystem simulation models. In T. J. Pitcher (Ed.), *Back to the Future: Advances in methodology for modelling and evaluating past ecosystems as future policy goals.* Fisheries Centre Research Reports 12, 94–102.

Cheung, W. W. L., & Sadovy, Y. (2004). Retrospective evaluation of data-limited fisheries: a case from Hong Kong. *Reviews in Fish Biology and Fisheries,* 14, 181–206.

Christensen, V., Guénette, S., Heymans, J. J., Walters, C. J., Watson, R., Zeller, D., & Pauly, D. (2003). Hundred-year decline of North Atlantic predatory fishes. *Fish and Fisheries,* 4, 1–24.

Christensen, V., & Pauly, D. (1992). ECOPATH II—An approach and a software for construction of ecosystem models and food web analysis. *ICES C.M.* 1992/L:30 (Poster).

Christensen, V., & Walters, C. J. (2004). Trade-offs in ecosystem-scale optimization of fisheries management policies. *Bulletin of Marine Science,* 74, 549–562.

CIHI. Canadian Institute for Health Information. (2004). *Improving the Health of Canadians.* Ontario: Canadian Institute for Health Information.

Clark, C. W. (1984). Strategies for multispecies management: Objectives and constraints. In R. M. May (Ed.), *Exploitation of marine communities* (pp. 302–312). Berlin: Springer-Verlag.

Clark, C. W. (1973). The economics of overexploitation. *Science,* 181, 630–634.

Coleman, J. S. (1988). Social capital in the creation of human capital. *American Journal of Sociology,* 94 (Supplement), S95–S120.

Coleman, F. C., Figueira, W. F., Ueland, J. S., & Crowder, L. B. (2004). The Impact of United States Recreational Fisheries on Marine Fish Populations. *Science,* 305, 1958–1960.

Committee on Animal Nutrition. (1993). *Nutrient requirements of fish*. Washington, DC: National Academy Press.

Committee on Food Protection, Food and Nutrition Board, National Research Council. (1973). *Toxicants occurring naturally in foods*. Washington, DC: National Academy of Sciences.

Connell, J. H., & Sousa, W. P. (1983). On the evidence needed to judge ecological stability or persistence. *The American Naturalist, 121*, 789–824.

Connors, M., Bisogni, C. A., Sobal, J., & Devine, C. M. (2001). Managing values in personal food systems. *Appetite, 36*, 189–200.

Constanza, R., d'Arge, R. de Groot, R., Farber, S., Grasso, M., Hannon, B., Limburg, K., Naeem, S., O'Neill, R. V., Paruelo, J., Raskin, R.G., Sutton, P., & van den Belt, M. (1997). The value of the world ecosystem services and natural capital. *Nature, 387*, 253–260.

Cooke, S. J., & Cowx, I. G. (2004). The role of recreational fishing in global fish crises *BioScience, 54*, 857–859.

Copeman, L. A., & Parrish, C. C. (2004). Lipid classes, fatty acids, and sterols in seafood from Gilbert Bay, Southern Labrador. *Journal of Agricultural and Food Chemistry, 52*, 4872–4881.

Copeman, L. A., & Parrish, C. C. (2003). Marine lipids in a cold coastal ecosystem: Gilbert Bay, Labrador. *Marine Biology, 143*, 1213–1227.

Counihan, C. M. (1999). *The anthropology of food and body*. London: Routledge.

Counihan, C. M., & Kaplan, S. L. (Eds.) (1998). *Food and gender: Identity and power*. Canada: Harwood Academic Publishers.

Counihan, C. M., & Van Esterik, P. (Eds.) (1997). *Food and culture: A reader*. New York: Routledge.

Courchamp, F., Clutton-Brock, T., & Grenfell, B. (1999). Inverse density dependence and the Allee effect. *Trends in Evolution and Ecology, 14*, 405–410.

Coward, H., Ommer, R. E., & Pitcher, T. J. (Eds.) (2000). *Just fish: Ethics and Canadian marine fisheries*. St. John's, NL: ISER Books.

Coyle, W., Hall, W., & Ballenger, N. (2001). Transportation technology and the rising share of U.S. perishable food trade. In A. Regmi, (Ed.), *Changing structure of food consumption and trade,* (pp. 31–40). Market and Trade Economics Division, Economic Research Service, U.S. department of Agriculture, Agriculture and Trade Report, WRS-01-1.

Crossan, J. D. (1996). *The historical Jesus*. Lansdowne Lecture, University of Victoria, February 12. (Videocassette, MacPherson Library, University of Victoria).

Cushing, D. H. (1987). *The provident sea*. Cambridge: Cambridge University Press.

Cuthbertson, D. P. (1947). *Report on nutrition in Newfoundland*. London: His Majesty's Stationary Office.

Cuthbertson, D. P. (1945). *Memo*, 17 July 1945. Provincial Archives of Newfoundland and Labrador. GN38 S 6-5-3 FILE 26.

Daily, G. C. (1997). *Nature's services: societal dependence on natural ecosystems*. Washington, DC: Island Press

Daskalov, G. M. (2002). Overfishing drives a trophic cascade in the Black Sea. *Marine Ecology Progress Series, 225*, 53–63.

Davies, I. M., & McKie, J. C. (1987). Accumulation of total tin and tributyltin in muscle tissue of farmed Atlantic salmon. *Marine Pollution Bulletin,* 18, 405–407.

Dayle, J., & McIntyre, L. (2003). Children's feeding programs in Atlantic Canada: Some Foucauldian theoretical concepts in action. *Social Science and Medicine,* 57, 313–325.

DeBoilieu, L. (1861/1969). *Recollections of Labrador life.* T.F. Bredin (Ed.), Toronto, ON: Ryerson Press.

Deimling, E., & Liss, W. J. (1994). Fishery development in the eastern North Pacific: a natural-cultural system perspective, 1888–1976. *Fisheries Oceanography,* 3(1), 60–77.

Deur, D. (2002). Rethinking precolonial plant cultivation on the Northwest coast of North America. *The Professional Geographer,* 54, 140–157.

Deur, D., & Turner, N. J. (Eds.) (2005). *Keeping it living. Traditions of plant use and cultivation on the Northwest Coast of North America.* Seattle, WA: University of Washington Press and Vancouver, BC: UBC Press.

Devine, J. A., Baker, K. D., & Haedrich, R. L. (2006). Deep-sea fishes qualify as endangered. *Nature,* 439, 29.

Diamond, J. (1997). *Guns, germs, and steel: The fates of human societies.* London, UK: Random House.

Dibsdall, L. A., Lambert, N., & Frewer, L. J. (2002). Using interpretative phenomenology to understand the food-related experiences and beliefs of a select group of low-income UK women. *Journal of Nutrition and Education Behaviour,* 34, 298–309.

Dickie, L. M., Kerr, S. R., & Boudreau, P. R. (1987). Size-dependent processes underlying regularities in ecosystem structure. *Ecological Monographs,* 57, 233–250.

Dickinson, R., & Leader, S. (1998). Ask the family. In S. Griffiths & J. Wallace (Eds.), *Consuming patterns: Food in the age of anxiety,* (pp. 122–129). Manchester, UK: Mandolin.

Dickson, B. (1997). From the Labrador Sea to global change. *Nature,* 386, 649–650.

Dietitians of Newfoundland and Labrador (the), Newfoundland and Labrador Public Health Association, and the Newfoundland and Labrador Association of Social workers. (2004). *ThecCost of Eating in Newfoundland and Labrador—2003.* 12 February. St. John's, NL.

Dorsa, D. (1995). The importance of ritual to children. *Dissertation Abstracts International,* 55, 12A. (AAT 9511405, 1995).

Dove, R. F. (1943). The diagnosis and treatment of the subclinical dietary deficiency diseases. *Northern Medical Review,* 1/1, 7–9.

Doyle, A. C. (1912). *The Lost World.* Toronto, ON: Musson.

Drake A., & Wilson, L. (1991). *Eulachon: a fish to cure humanity.* Museum Note No. 32. Vancouver, BC: UBC Museum of Anthropology.

Drewnowski, A. (1997). Taste preferences and food intake. *Annual Review of Nutrition,* 17, 237–253.

Drucker, P. (1951). *The Northern and Central Nootkan tribes.* Bureau of American Ethnology, Bulletin No. 144. Washington D.C.: Smithsonian Institution.

Duarte, C. M. (2002). The future of seagrass meadows. *Environmental Conservation,* 29, 192–206.

Duarte, C. M. (1989). Temporal biomass variability and production/biomass relationships of seagrass communities. *Marine Ecology Progress Series,* 51, 269–276.

Dugan, J. E., & Davis, G. E. (1993). Applications of marine refugia to coastal fisheries management. *Canadian Journal of Fisheries and Aquatic Sciences,* 50, 2029–2042.

Dunn, J. R. (1979). Predatory-prey interactions in the Eastern Bering Sea. In H. Clepper (Ed.), *Predator-Prey Systems in Fisheries Management.* Washington, DC: Sport Fishing Institute.

EAQ, *Education Administration Quarterly.* (2004). Special issue: Social justice challenges to educational administration. *Educational Administration Quarterly*, 40 (1).

Ecotrust. (2004). *Catch 22: Conservation, communities and the privatization of BC fisheries.* Vancouver, BC and Portland, OR: Ecotrust.

Edge Research. (2002). *Public opinion poll on fully protected marine areas in New England and Atlantic Canada.* Arlington, VA: Edge Research. *http://www.environmentaldefense. org/documents/1532_MPApoll.pdf*

Elliot, R. D. 1991. The management of the Newfoundland Turr hunt. In A. J. Gaston & R. D. Elliot (Eds.), Studies of high-latitude seabirds 2: Conservation biology of thick-billed Murres in the Northwest Atlantic. *Canadian Wildlife Service Occasional Paper,* 69, 29–35.

Elliot, R. D., Collins, B. T., Hayakawa, E. G., & Métras, L. (1991). The harvest of murres in Newfoundland from 1977-78 to 1987-88. In A. J. Gaston & R. D. Elliot (Eds.), *Studies of High-Latitude Seabirds* 2: Conservation biology of thick-billed Murres in the Northwest Atlantic (pp. 36–44). Canadian Wildlife Service Occasional Paper 69.

Elton, C. S. (1942). *Voles, mice and lemmings.* Oxford: Clarendon Press.

Elton, C. S. (1927). *Animal ecology.* London: Sidgwick and Jackson.

Ervik, A., Hansen, P. K, Aure, J., Stigebrandt, A., Johannessen, P., & Jahnsen, T. (1997). Regulating the local environmental impact of intensive marine fish farming I. The concept of the MOM system (Modelling—Ongrowing fish farms—Monitoring). *Aquaculture*, 158, 85–94.

Estes, J. A., Tinker, M. T., Williams, T. M., & Doak, D. F. (1998). Killer whale predation on sea otters linking oceanic and nearshore ecosystems. *Science,* 282, 473–476.

Eythorsson, E. (1998). Voices of the weak—Relational aspects of local ecological knowledge in the fisheries. In S. Jentoft (Ed.), *Commons in a Cold Climate: Coastal fisheries and reindeer pastoralism in North Norway: The co-management approach* (pp. 185–204). Paris and New York: UNESCO and Parthenon Publishing Group.

FAO, Food and Agriculture Organization of the United Nations. (2000). *The state of world fisheries and aquaculture* (SOFIA). FAO Information Division.

FAO, Food and Agriculture Organization of the United Nations. (1995). *Code of conduct for responsible fisheries.* Rome, Italy: FAO.

Felt, L. F., Murphy, K., & Sinclair, P. R. (1995). Everyone does it: unpaid work and household reproduction. In L. F. Felt & P. R. Sinclair, (Eds.), *Living on the edge: the Great Northern Peninsula of Newfoundland* (pp. 77–102). St. John's: ISER Books.

Felt, L. F., & Sinclair, P. R. (Eds.) (1995). *Living on the edge: The Great Northern Peninsula of Newfoundland.* St. John's, NL: ISER Books.

Felt, L. F., & Sinclair, P. R. (1992). Everyone does it: unpaid work in a rural peripheral region. *Work, Employment and Society,* 6(1), 43–64.

Finlayson, A. C. (1994). *Fishing for truth.* St. John's, NL: ISER Press.

Fischer, J., Haedrich, R. L., & Sinclair, P. R. (1997). *Interecosystem impacts of forage fish fisheries.* Forage Fishes in Marine Ecosystems. Proceedings of the International

Symposium on the Role of forage Fishes in Marine Ecosystems. Alaska Sea Grant College Program Report No. 97-01. University of Alaska Fairbanks.

Fisheries and Oceans Canada. (2005). *Canadian aquaculture production statistics.* Statistics Canada - Cat. no. 23-222-XI E.*http://www.dfo-mpo.gc.ca/communic/statistics/aqua/ aqua00_e.htm.*

Fodor, J. G., Abbott, E. D., & Rusted, I. E. (1973). An epidemiological study of hypertension in Newfoundland. *Canadian Medical Association Journal,* 108, 1365–1368.

Folke, C., Perrings, C., McNeely, J. A., & Myers, N. (1993). Biodiversity conservation with a human face: Ecology, economics and policy. *AMBIO,* 22 (2–3), 62–63.

Forbush, E. H. (1912). *Game birds, wild-fowl and shorebirds of Massachusetts.* Boston, MA: Massachusetss Board of Agriculture.

Forster, J. (1978). *Working for wildlife: The beginning of preservation in Canada.* Toronto, ON: University of Toronto Press.

Foster, W. (2004). The decline of the local: A challenge to educational leadership. *Educational Administration Quarterly,* 40, 176–191.

Fraser A. J., & Sargent, J. R. (1989). Formation and transfer of fatty acids in an enclosed marine food chain comprising phytoplankton, zooplankton and herring (*Clupea harengus* L.) larvae. *Marine Chemistry* 27, 1–18.

Freire, P. (1998). *Teachers as cultural workers.* Boulder, CO: Westview Press.

Freuchen, P., & Salomonsen, F. (1958). *The Arctic year.* New York, NY: Putnam.

Friscolanti, M. (2003b, October 27). Poor diet leads to big trouble for Newfoundlanders. *National Post.*

Friscolanti, M. (2003a, October 27). Teen heart disease coming: expert. *National Post,* p. 1.

Fuller, S., Fuller, C., & Cohen, M. (2003). *Health care restructuring in BC.* Vancouver: Canadian Centre for Policy Alternatives.

Gamberg, M. (1997). *Contaminants in Yukon moose and caribou-1996.* A Report prepared for Yukon Contaminants Committee and Department of Indian and Northern Affairs. Northern Contaminants Program. Whitehorse, Yukon.

García, S. M. (1997). Fisheries management and sustainability: A new perspective of an old problem. In D. A. Hancock, D. C. Smith, A. Grant & J. P. Beumer (Eds.), *Developing and sustaining world fisheries resources. The state of science and management* (pp. 631–654). Second World Fisheries Congress, Brisbane, Australia, 28 July – 2 August 1996. Melbourne: CSIRO Publishing.

Garibaldi, A., & Turner, N. (2004). *Cultural keystone species: implications for ecological conservation and restoration. Ecology and Society,* 9(3), 1. *http://www. ecologyandsociety.org/vol9/iss3/art1*

Gasciogne, J., & Lipcius, R. N. (2004). Allee effects in marine systems. *Marine Ecology Progress Series,* 269, 49–59.

Gedalof, Z., & Smith, D. J. (2001). Interdecadal climate variability and regime-scale shifts in Pacific North America. *Geophysical Research Letters,* 28, 1515–1518.

Gehlar, M., & Coyle, W. (2001). Global food consumption and impacts on trade patterns. In A. Regmi, (Ed.), *Changing structure of food consumption and trade,* (pp. 4–13). Market and Trade Economics Division, Economic Research Service, U.S. department of Agriculture, Agriculture and Trade Report, WRS-01-1.

George, S. (1999). *A short history of neo-liberalism: Twenty years of elite economics and emerging opportunities for structural change.* Paper presented at the Conference on Economic Sovereignty in a Globalising World, March 24–26, Bangkok.

Gibson, R. J., & Haedrich, R. L. (1988). The exceptional growth of juvenile Atlantic salmon, *Salmo salar*, in the city waters of St. John's, Newfoundland, Canada. *Polish Archives of Hydrobiology* 35, 385–407.

Gibson, J., Haedrich, R., Kennedy, J. C., Vodden, K., & Luther, R. J. (Submitted). Promoting, blocking and diverting the flow of knowledge: Four case studies from Newfoundland and Labrador. In J. Lutz & B. Neis (Eds.), *Making and moving knowledge.* Montreal: McGill–Queens University Press.

Gilchrist, G., Mallory, M., & Merkel, F. (2005). Can local ecological knowledge contribute to wildlife management? Case studies of migratory birds. *Ecology and Society,* 10, 20. {online] URL: *http://www.ecologyandsociety.org/vol10/iss1/art20/.*

Giroux, H., & Aronowitz, S. (1985). *Education under siege: The conservative, liberal, and radical debate over schooling.* South Hadley: Bergin and Garvey.

Gislason, H., Sinclair, M., Sainsbury, K., & O'Boyle, R. (2000). Symposium overview: incorporating ecosystem objectives within fisheries management. *ICES Journal of Marine Science,* 57, 468–475.

Gislason, G., Lam, E., & Mohan, M. (1996). Fishing for Answers: Coastal Communities and the BC Salmon Fishery. Final Report prepared for BC Job Protection Commission. Vancouver, BC: The Ara Consulting Group.

Glavin, T. (1996). *Dead reckoning: Confronting the crisis in Pacific fisheries.* Vancouver BC: The David Suzuki Foundation, and Toronto, ON: Greystone Books, Douglas and McIntyre.

Goldsmith, G. (1950). Relationships between nutrition and pregnancy as observed in recent studies in Newfoundland. *American Journal of Public Health,* 40, 953–959.

Goldsmith, G., Darby, W.J., Steinkamp, R.C., Stockwell Beam, A., & McDevitt, E. (1950). Resurvey of nutritional status in Norris Point Newfoundland. *Journal of Nutrition,* 40, 41–70.

Gollop, J. B., Barry, T. W., & Iverson, E. H. (1986). *Eskimo curlew: A vanishing species?* Regina, SK: Saskatchewan Natural History Society.

Gomes, M. C. (1993). *Predictions under uncertainty. Fish assemblages and food webs on the Grand Banks of Newfoundland.* St. John's, NL: ISER Books.

Goodman, J. (1995). Change without difference: School restructuring in historical perspective. *Harvard Educational Review,* 65, 1–29.

Goodnight, S. H. (1996). The fish oil puzzle. *Science and Medicine*, September/October, 42–51.

Gosling, W. G. (1910). *Labrador: Its discovery, exploration, and development.* London: Alston Rivers.

Gosse, K. R., & Wroblewski, J. S. (2004). Variant colourations of Atlantic cod (*Gadus morhua*) in Newfoundland and Labrador nearshore waters. *ICES Journal of Marine Science,* 61, 752–759.

Goudie, R. I., Robertson, G. J., & Reed, A. (2000). Common Eider (*Somateria mollissima*). In A. Poole & F. Gill (Eds.), *Birds of North America* (No. 546) Philadelphia, PA: American Ornithologists' Union.

Goudzwaard, B., & DeLange, H. (1995). *Beyond poverty and affluence: Towards a Canadian economy of care.* Toronto, ON: University of Toronto Press.

Government of Canada. (1992). *Canada's food guide to healthy living.* Ottawa, ON: Government of Canada.

Grande, J. P., & Donadio, J. V. (1998). Dietary fish oil supplementation in IgA nephropathy: A therapy in search of a mechanism? *Nutrition,* 14, 240–242.

Gray-Donald, K., Jacobs-Starkey, L., & Johnson-Down, L. (2000). Food habits of Canadians: Reduction in fat intake over a generation. *Canadian Journal of Public Health*, 91, 381–385.

Great Britain. (1939). Economic Advisory Council, Committee on Nutrition in the British Empire. "First Report—Part I: Nutrition in the Colonial Empire." CMD 6050. London.

Green, J. M., & Wroblewski, J. S. (2000). Movement patterns of Atlantic cod in Gilbert Bay, Labrador: evidence for bay residency and spawning site fidelity. *Journal of Marine Biology*, 80, 1077–1085.

Greenberg, R., & Reaser, J. (1995). *Bring back the birds.* Mechanicsburg, PA: Stackpole.

Gregg, A. (2005). Quebec's final victory. *The Walrus,* 2 (1), 50–61.

Groscolas, R. (1993). De l'huile pour traiter l'obésité? *La Recherche,* 24, 1404–1406.

Guénette, S., Pitcher, T. J., & Walters, C. J. (2000). The potential of marine reserves for the management of northern cod in Newfoundland. *Bulletin of Marine Science*, 66, 831–852.

Guggenheim, K. (1981) *Nutrition and nutritional diseases: The evolution of concepts.* Toronto, ON: DC Heath.

Gunther, F. (2001). Fossil energy and food security. *Energy and Environment,* 12, 253–273.

Habermas, J. (1971). *Knowledge and human interests*, trans. Shapiro. Boston, MA: Beacon.

Haedrich, R. L. (1997). Distribution and population ecology. In D. J. Randall & A. P. Farrell (Eds.), *Deep-Sea Fishes* (pp. 79–114). London: Academic Press.

Haedrich, R. L., & Hamilton, L. C. (2000). The fall and future of Newfoundland's cod fishery. *Society and Natural Resources,* 13, 359–372.

Haedrich, R. L., Merrett, N. R., & O'Dea, N. R. (2001). Can ecological knowledge catch up with deep-water fishing? A North Atlantic perspective. *Fisheries Research,* 51, 113–122.

Haggan, N. (2000). Back to the Future and creative justice: Recalling and restoring forgotten abundance in Canada's marine ecosystems. In H. Coward, R. E. Ommer & T. J. Pitcher (Eds). *Just fish: Ethics in the Canadian coastal fisheries* (pp. 83–99). St. John's, NL: ISER Books.

Haggan, N., & Brown, P. (2003). Aboriginal fisheries issues: the West Coast of Canada as a case study. In D. Pauly & M. L. D Palomares (Eds.), *Production systems in fishery Management* (pp. 17–19). UBC Fisheries Centre Research Reports, 10(8), 28p.

Haggan, N., Turner, N.J., Carpenter, J., Jones, J.T., Menzies, C., & Mackie, Q. (2006). 12,000+ years of change: Linking traditional and modern ecosystem science in the Pacific Northwest. Fisheries Centre Working Paper #2006-02, The University of British Columbia, Vancouver, British Columbia, Canada. *http://www.fisheries.ubc.ca /publications/working/2006/2006-02*.pdf last accessed July 2, 2006.

Hahm, W., & Langton, R. (1984). Prey selection based on predator/prey weight ratios for some northwest Atlantic fish. *Marine Ecology Progress Series,* 19, 1–5.

Haley, M. (2001). Changing consumer demand for meat: The U.S. example, 1970–2000. In A. Regmi, (Ed.), *Changing structure of food consumption and trade*, (pp. 41–48). Market and Trade Economics Division, Economic Research Service, U.S. department of Agriculture, Agriculture and Trade Report, WRS-01-1.

Hall, P. O. J., Anderson, L. G., Holby, O., Kollberg, S., & Samuelsson, M. O. (1990). Chemical fluxes and mass balances in a marine fish cage farm. *Marine Ecology Progress Series,* 61, 61–73.

Hall, S. J. (1999). *The effects of fishing on marine ecosystems and fisheries.* Oxford: Blackwell.

Hallaq, H., & Leaf, A. (1992). Stabilization of cardiac arrhythmias by ω-3 polyunsaturated fatty acids. In A. Sinclair & R. Gibson (Eds.), *Essential fatty acids and Eicosanoids* (pp. 245–247). Champaign, IL: American Oil Chemists' Society.

Hallin, L. (2000). *A guide to the BC economy and labour market.* Victoria, BC: Ministry of Finance and Corporate Relation.

Halter, M. (2000). *Shopping for identity: The marketing of ethnicity.* New York: Schocken Books.

Hamelin, A. M., Habicht, J. P., & Beaudry, M. (1999). Food insecurity: Consequences for the household and broader social implications. *Journal of Nutrition,* 129 (Suppl. 2), 525S–528S.

Hanrahan, M., & Ewtushik, M. (2001). *A veritable scoff: Sources of foodways and nutrition in Newfoundland and Labrador.* St. John's, NL: Flanker Press.

Hansen, P. K., Ervik, A., Schaaning, M., Johannessen, P., Aure, J., Jahnsen, T., & Stigebrandt, A. (2001). Regulating the local environmental impact of intensive marine fish farming II. The monitoring programme of the MOM system (Modelling—Ongrowing fish farms—Monitoring) *Aquaculture,* 194, 75–92.

Hardin, G. 1968. The tragedy of the commons. In A. Markandya & J. Richardson (Eds.), *The Earthscan reader in Environmental Economics,* London: Earthscan. 1992.

Harper, C. L., & Le Beau, B. F. (2003). *Food society and the environment.* New Jersey: Prentice Hall.

Harris, C. E. (2004). Communication technologies and leadership for resilience: Participatory research outreach in five coastal communities. In *Education and Social Action Conference Proceedings* (pp. 204–208). Sydney, Australia: University of Technology.

Harris, C. E., Riley, S., & Robinson, L. (Forthcoming). A question of power: Linking political will, discourse and health in a coastal community school. In R. E. Ommer & P. R. Sinclair (Eds.), *Power and Restructuring: Canada's Coastal Society and Environment.* St. John's, NL: ISER Press.

Harris, D. D. (2005). Indian Reserves, Aboriginal Fisheries, and the public's right to fish in British Columbia, 1876–82., in: John McLaren A.R. Buck, and Nancy E. Wright (Eds.), *Despotic Dominion. Property Rights in British Settler Societies* (pp. 266–293). Vancouver, BC: UBC Press.

Harris, L. (1990). *Independent review of the state of Northern Cod.* Ottawa, ON: Supply and Services Canada.

Harris, M. (1998). *Lament for an ocean. The collapse of the Atlantic cod fishery. A true crime story.* Toronto, ON: McClelland and Stewart Inc.

Harris, R. C., & Warkentin, J. (1974). *Canada before confederation: a study in historical geography.* New York: Oxford University Press.

Harris, W. S. (1997). n-3 fatty acids and serum lipoproteins: human studies. *American Journal of Clinical Nutrition,* 65, 1645–1654.

Harrison, T. W., & Kachur, J. L. (Eds.) (1999). *Contested classrooms: Education, globalization and democracy in Alberta.* Edmonton, AB: University of Alberta Press/Parkland Institute.

Hastings, A. (1988). Food web theory and stability. *Ecology,* 69, 1665–1668.

Hatab, L. J. (2002). Heidegger and the question of empathy. In F. Raffoul & D. Pettigrew (Eds.), *Heidegger and practical philosophy* (pp. 249–272). New York: SUNY Press.

Haury, L. R., McGowan, J. A., & Wiebe, P. H. (1978). Patterns and processes in the time-space scales of plankton distributions. In J. H. Steele (Ed.), *Spatial Pattern in Plankton Communities* (277–327). New York: Plenum Press.

Hauser, L., Adcock, G. J., Smith, P. J., Bernal Ramírez, J. H., & Carvalho, G.R. (2002). Loss of microsatellite diversity and low effective population size in an overexploited population of New Zealand snapper (*Pagrus auratus*). *Proceedings of the National Academy of Sciences, USA,* 99(18), 11742–11747.

Hay, D. I. (2000). *School-based feeding programs: A good choice for our children?* Victoria, BC: Information Partnership.

Hays, G. C., Broderick, A. C., Godley, B. J., Luschi. P., & Nichols, W. J. (2003). Satellite telemetry suggests high levels of fishing-induced mortality in marine turtles. *Marine Ecology Progress Series,* 262, 305–309.

Health Canada. (2002). *Canada's food guides from 1942 to 1992.* Catalogue H39-651/2002E-IN. Ottawa, ON: Health Canada.

Health and Welfare Canada. (1990). *Action towards healthy eating: Technical report. Reports of the Task Group and Technical Group on Canada's Food Guide and the Task Group on Food Consumption to the Communications/Implementation Committee.* Ottawa.

Heidegger, M. (1977). *The question concerning technology and other essays,* trans. William Lovitt. New York, NY: Harper and Row.

Henderson, B. A. (1987). Interspecific relations among fish species in South Bay, Lake Huron, 1949–84. *Canadian Journal of Fisheries and Aquatic Sciences,* 44 (Suppl.2), 10–14.

Henderson, R. J., Forrest, D. A. M., Black, K. D., & Park, M. T. (1997). The lipid composition of sealoch sediments underlying salmon cages. *Aquaculture,* 158, 69–83.

Henry, P. (1945). A whole nation is sick. *Magazine Digest,* October, 73–77.

Hertzman, C., & Siddiqi, A. (2000). Health and rapid economic change in the late twentieth century. *Social Science and Medicine,* 51, 809–819.

Heymans, J. J. (2003). First Nations impact on the Newfoundland ecosystem during pre-contact times. In J. J. Heymans (Ed.), *Ecosystem models of Newfoundland and Southeastern Labrador: Additional information and analyses for 'Back to the Future'.* Fisheries Centre Research Report, 11(5), 4–12.

Heymans, J. J., & Pitcher, T. J. (2004). Synoptic methods for constructing models of the past. In T. J. Pitcher (Ed.), *Back to the Future: Advances in methodology for modelling and evaluating past ecosystems as future policy goals* (pp. 11–17). UBC Fisheries Centre Research Reports, 12(1): 158 pp.

Hicks, K. B., & Moreau, R. A. (2001). Phytosterols and phytostanols: functional food cholesterol busters. *Food Technology,* 55, 63–67.

Hilborn, R., Quinn, T. P., Schindler, D. E., & Rogers, D. E. (2003). Biocomplexity and fisheries sustainability. *Proceedings of the National Academy of Sciences, USA,* 100(11), 6564–6568.

Hilborn, R., & Walters, C. J. (1992). *Quantitative fisheries stock assessment: choice, dynamics and uncertainty*. New York, NY: Chapman and Hall.

Hites, R. A., Foran, J. A., Carpenter, D. O., Hamilton, M. C., Knuth, B. A., & Schwager, S. J. (2004). Global assessment of organic contaminants in farmed salmon. *Science, 303*, 226–229.

Hood, R. J., & Fox, Ben (Producers and Directors). 2003. *Gitga'ata Spring Harvest—Traditional knowledge of Kiel*. A Co-production by the Gitga'at Nation and *Coasts Under Stress* major collaborative research initiative, University of Victoria, BC.

Howley, J. P. (1915). *The Beothuks or Red Indians. The Aboriginal inhabitants of Newfoundland*. Cambridge, U.K.: Cambridge University Press.

Hueting, R., & L. Reijndiers. (1998). Sustainability is an objective concept. *Ecological Economics, 27*, 139–147.

Huizinga, J. (1971). *Homo ludens: A study of the play element in culture*. London, UK: Granada Publishing.

Hulse, J. H. (1995). *Science, agriculture, and food security*. Ottawa, ON: NRC Research Press.

Hutchings, J. A., & Myers, R. A. (1994). What can be learned from the collapse of a renewable resource? Atlantic cod, *Gadus morhua*, of Newfoundland and Labrador. *Canadian Journal of Fisheries and Aquatic Sciences, 51*, 2126–2146.

Huxley, T. H. (1883) *'Inaugural address'*, International Fisheries Exhibition, London Vol. 4, pp. 1–22.

ICES, International Council for the Exploration of the Sea. (1995). Herring assessment working group for the area south of 62^0N. *ICES C.M.* 1995/Assess: 13. Copenhagen, Denmark: ICES.

Ife, Jim. (2002). *Community development: Community-based alternatives in an age of globalization* (2nd ed) . French Forest, NSW, Australia: Pearson Education.

IFFO, International Fishmeal and Fish Oil Organisation. (n.d.). Industry Overview: Usage by region. http://www.iffo.net/default.asp?fname=1andsWebIdiomas=1andurl=253.

Innis, H. (1954). *The cod fishery. The history of an international economy* (Rev. ed.). Toronto, ON: University of Toronto Press.

Isaacs, J. C. (2000). The limited potential of ecotourism to contribute to wildlife conservation. *Wildlife Society Bulletin, 28*, 61–69.

Iwama, G. K. (1991). Interactions between aquaculture and the environment. *Critical Reviews in Environmental Control, 21*, 177–216.

Jackson, J. B. C., Kirby, M. X., Berger, W.H., Bjorndal, K.A., Botsford, L. W., Bourque, B.J., *et al.* (2001). Historical overfishing and the recent collapse of coastal ecosystems. *Science, 293*, 629–637.

Jacobs M. N., Covaci, A., & Schepens, P. (2004). Time trend investigation of PCBs, PDBEs, and organochlorine pesticides in selected *n*–3 polyunsaturated fatty acid rich dietary fish oil and vegetable oil supplements; nutritional relevance for human essential *n*–3 fatty acid requirements. *Journal of Agricultural and Food Chemistry, 52*, 1780–1788.

Jacobs, M. N., Covaci, A., & Schepens, P. (2002). Investigation of selected persistent organic pollutants in farmed Atlantic salmon (*Salmo salar*), salmon aquaculture feed, and fish oil components of the feed. *Environmental Science and Technology, 36*, 2797–2805.

Jenkins, E. W. (2003). Environmental education and the public understanding of science. *Frontiers in Ecology and the Environment, 1*, 437–433.

Jenness, D. (1965). *Eskimo Administration: III Labrador*. Montreal: Arctic Institute of North America, Technical Paper No. 16.

Jennings, S., Pinnegar, J. K., Polunin, N. V. C., & Warr, K. J. (2002). Linking size-based and trophic analyses of benthic community structure. *Marine Ecology Progress Series, 226,* 77–85.

Jennings, S., Reynolds, J. D., & Mills, S. C. (1998). Life history correlates of responses to fisheries exploitation. *Proceedings of the Royal Society of London, Series B*, 265, 333–339.

Johns, T. (1996). *The origins of human diet and medicine: Chemical ecology*. Tucson, AZ: University of Arizona Press.

Johns, T., & Eyzaguirre, P. (2003). *Dietary diversity: Linking biodiversity conservation to new challenges in global health*. Poster for Biodiversity and Health Conference, Ottawa, ON, Montreal, QC: Centre for Indigenous Peoples' Nutrition and Environment; McGill University and Rome, Italy: International Plant Genetic Resources Institute.

Johns, T., & Romeo, R. T. (Ed.) (1996). Functionality of food phytochemicals. Meeting of the Phytochemical Society of North America on Functionality. *Recent Advances in Phytochemistry* Vol. 31, New York And London: Plenum Press.

Johnsen, R. I., Grahl-Nielsen, O., & Lunestad, B. T. (1993). Environmental distribution of organic waste from a marine fish farm. *Aquaculture,* 118, 229–244.

Johnson, G. (n.d.). *Nutritional deficiency diseases in Newfoundland and Labrador: Their recognition and elimination*. Typescript, 34 pp.

Jones, J. T. (2002)."*We looked after all the salmon streams." Traditional Heiltsuk Cultural Stewardship of Salmon and Salmon Streams: A Preliminary Assessment*. Masters Thesis, Environmental Studies, University of Victoria.

Jones, R. R. (1999). Haida names and utilization of common fish and marine mammals. In N. Haggan & A. Beattie (Eds.), *Back to the Future: A Re-construction of the Hecate Strait Ecosystem as it might have been in the early 1900s* (pp. 39–48). UBC Fisheries Centre Research Reports, (7), 3.

Jones, R. R., & Williams-Davidson. T. (2000). Applying Haida ethics in today's fishery. In H. Coward, R. E. Ommer & T. J. Pitcher (Eds.), *Just fish: Ethics and Canadian marine fisheries* (pp. 100–117). St. John's, NL: ISER Books.

Josselson, R. (1994). The theory of identity development and the question of intervention: An introduction." In S. L. Archer (Eds.), *Interventions for adolescent identity development* (pp. 12–25). Thousand Oaks, CA: Sage.

Justice, C. L. (2000). *Mr Menzies' garden legacy. Plant collecting on the Northwest Coast*. Vancouver, BC: Cavendish Books.

Kalantzi, O. I., Alcock, R. E., Johnston, P. A, Santillo, D., Stringer, R. L., Thomas, G. O., & Jones, K. C. (2001). The global distribution of PCBs and organochlorine pesticides in butter. *Environmental Science and Technology,* 35, 1013–1018.

Karst, A. (2005). *The ethnoecology and reproductive ecology of bakeapple (*Rubus chamaemorus *L., Rosaceae) in Southern Labrador*. MSc Thesis, Biology and Environmental Studies, University of Victoria.

Keane, A. (1992). Too hard to swallow? The palatability of healthy eating advice. In M. Mills (Ed.). *The politics of dietary change*, (pp. 172–192). Aldershot, England: Dartmouth Publishing Company.

Keating, D. P., & Hertzman, C. (Eds.) (1999). *Developmental health and the wealth of nations: Social, biological and educational dynamics*. New York, NY: Guilford Press.

Kennedy, J. C. (Forthcoming). Disempowerment: The cod moratorium, fisheries restructuring and the decline of power among Labrador fishers. In P. R. Sinclair & R. E. Ommer (Eds.), *Power, agency and nature: Coastal society and environment* (Chapter 6). St. John's, NL: ISER Books.

Kennedy, J. C. (1997). Labrador Metis Ethnogenesis. *Ethnos*, 3–4, 5–23.

Kennedy, J. C. (1996). *Labrador village*. Prospect Heights, IL: Waveland Press.

Kennedy, J. C. (1995). *People of the bays and headlands: Anthropological history and the fate of communities in the unknown Labrador*. Toronto, ON: University of Toronto Press.

Kennedy, J. C. (1982). Holding the Line: Ethnic boundaries in a Northern Labrador community. St. John's: ISER Social and economic studies, No. 27.

Kerstetter, S. (2003). *Rags and riches: Wealth inequality in Canada*. Ottawa, ON: CCPA.

Khan, M. A., Parrish, C. C., & Shahidi, F. (2004). Microbial quality of cultured Newfoundland mussels and scallops. In F. Shahidi & B. K. Simpson (Eds.), *Seafood Quality and Safety: Advances in the New Millennium* (pp. 317–325). St. John's, NL: ScienceTech Publishing Company.

Keleher, J. J. (1972). Great Slave Lake: Effect of exploitation on the salmonid community. *Journal of the Fisheries Research Board of Canada*, 29, 741–753.

Kim, C., McKay, F., Kuhnlein, H. V., Receveur, O. and Chan, H. M. (1994). *Dietary intake of cadmium from traditional food in Fort Resolution, Northwest Territories*. Montreal, QC: The 27th Annual Symposium of the Society of Toxicology of Canada.

Kitchen Corner. (1947–68). *Provincial Archives of Newfoundland and Labrador*. GN 26/12 Box A and Box B.

Klein, S., & Long, A. (2003). *A bad time to be poor: An analysis of British Columbia's new welfare policies*. Vancouver, BC: CCPA.

Kleivan, H. (1966). *The Eskimos of Northeast Labrador: A History of Eskimo–White Relations 1771–1955*. Oslo, Norway: Norsk Polarinstitutt.

Koehne, M. (1922). Cooperative child nutrition service. *Journal of Home Economics*, 14.11, 535–544.

Krzynowek, J. (1985). Sterols and fatty acids in seafood. *Food Technology*, February, 61–68.

Kuhnlein, H. V. (1992). Change in the use of traditional food by the Nuxalk Native people of British Columbia. In G. H. Pelto & L. A. Vargas (Eds.), *Perspectives on dietary change. Studies in nutrition and society*, Special Issue, *Ecology of Food and Nutrition*, 27, 259–282.

Kuhnlein, H. V. (1989). Factors influencing use of traditional foods among the Nuxalk people. *Journal of the Canadian Dietetic Association*, 50, 102–108.

Kuhnlein, H. V. (1984). Traditional and contemporary Nuxalk foods. *Nutrition Research*, 4, 789–809.

Kuhnlein, H. V., Chan, A. C., Thompson, J. N., & Nakai, S. (1982). Ooligan grease: A nutritious fat used by Native people of Coastal British Columbia. *Journal of Ethnobiology*, 2(2), 154–161.

Kuhnlein H. V., & Chan, H. M. (2000). Environment and contaminants in traditional food systems of Northern Indigenous Peoples. *Annual Review of Nutrition*, 20, 595–626.

Kuhnlein, H. V., & Receveur, O. (1996). Dietary change and traditional food systems of Indigenous Peoples. *Annual Review of Nutrition*, 16, 417–442.

Kuhnlein, H. V.. & Turner, N. J. (1991). *Traditional plant foods of Canadian Indigenous People. Nutrition, botany and use*. Philadelphia, PA: Gordon and Breach Science Publishers.

Kuhnlein H. V., Yeboah, F., Sedgemore, M., Sedgemore, S., & Chan, H. M. (1996). Nutritional qualities of ooligan grease: a traditional food fat of British Columbia First Nations. *Journal of Food Composition and Analysis, 9*, 18–31.

Kurlansky, M. (1997). *Cod, a biography of the fish that changed the world*. Toronto, ON: A.A. Knopf Canada.

Krebs-Smith, S. M., Heimendinger, J., Patterson, B. H., Subar, A. F., Kessler, R., & Pivonka, E. (1995). Psychosocial factors associated with fruit and vegetable consumption. *American Journal of Health Promotion, 10*, 98–104.

Lai, H-L., &. Gallucci, V. F. (1988). Effects of parameter variability on length-cohort analysis. *Journal du Conseil international pour l'exploration de la mer, 45*, 82–92.

Lancet. (1945). Nutrition in Newfoundland. June 16, 760–761.

Large, P. A., Lorance, P., & Pope, J. G. (1998). The survey estimates of the overall size composition of the deepwater fish species on the European continental slope, before and after exploitation. *ICES CM, 1998/O*, 24.

Larkin, P. A. (1996). Concepts and issues in marine ecosystem management. *Reviews in Fish Biology and Fisheries, 6*, 139–164.

Law, R. (2000). Fishing, selection, and phenotypic evolution. *ICES Journal of Marine Science, 57*, 659–668.

Lawson J. W., J. T. Anderson, E. L. Dalley, and G. B. Stenson (1998). Selective foraging by harp seals *Phoc groenlandica* in nearshore and offshore waters on Newfoundland, 1993 and 1994. *Marine Ecology Progress Series, 163*, 1–10.

Lear, W. H. (1998). History of fisheries in the Northwest Atlantic: The 500-year perspective. *Journal of Northwest Atlantic Fisheries Sciences, 23*, 41–71.

Lepofsky, D., Turner, N. J., & Kuhnlein, H. V. (1985). Determining the availability of traditional wild plant foods: An example of Nuxalk foods, Bella Coola, British Columbia. *Ecology of Food and Nutrition, 16*, 223–241.

Lerner, R. M., Taylor, C. S., & von Eye, A. (2002). Positive youth development: Thriving as the basis of personhood and civil society. In R. Lerner, C. Taylor & A. von Eye (Eds.), *New directions for Youth Development: Pathways to positive development among diverse youth* (pp. 11–34). San Francisco, CA: Jossey-Bass.

Levin, L. A., Boesch, D. F., Covich, A., Dahm, C., Erseus, C., Ewel, K. C., Kneib, R. T., Moldenke, A., Palmer, M. A., Snelgrove, P., Strayer, D., & Weslawski, J. M. (2001). The function of marine critical transition zones and the importance of sediment biodiversity. *Ecosystems, 4*, 430–451.

Lien, J., Fawcett, K., & Staniforth, S. (1985). *Wet and fat: Whales and seals of Newfoundland and Labrador*. St. John's, NL: Breakwater.

Lilly, G. R. (1994). Predation by Atlantic cod on capelin on the southern Labrador and Northeast Newfoundland shelves during a period of changing spatial distributions. *ICES Marine Science Symposia, 198*, 600–611.

Lilly, G. R. (1991). Interannual variability in predation by cod (*Gadus morhua*) on capelin (*Mallotus villosus*) and other prey off southern Labrador and northeastern Newfoundland. *ICES Marine Science Symposia, 193*, 133–146.

Lilly, G. R. (1987). Interactions between Atlantic cod (*Gadus morhua*) and capelin (*Mallotus villosus*) off Labrador and eastern Newfoundland: a review. *Canadian Technical Report of Fisheries and Aquatic Sciences* No. 1567.

Lilly, G. R.; Hop, H., Stansbury, D. E., & Bishop, C. A. (1994). Distribution and abundance of polar cod (*Boreogadus saida*) of Southern Labrador and Eastern Newfoundland. *ICES Council Metting Papers*, 1994 No. ICES-CM-1994/0:6, 22 pages.

Little, J. M. (1914). Beriberi. *Journal of the American Medical Association*, LXIII, 13, 1288.

Little, J. M. (1912). Beriberi caused by fine white flour. *Journal of the American Medical Association,* LVIII, 28, 2030.

Longhurst, A. (1999). Double vision: two views of the North Atlantic fishery. *Fisheries Research,* 40 (1), 1–5.

Ludwig, D., Hilborn, R., & Walters, C. (1993). Uncertainty, resource exploitation and conservation: Lessons from history. *Science,* 260, 17–36.

Lund, T. A. (1980). *American wildlife law*. Berkeley, CA: University of California.

Lundvall, D., Svanback, R. Persson, L., & Bystrom. P. (1999). Size-dependent predation in piscivores: interactions between predator foraging and prey avoidance abilities. *Canadian Journal of Fisheries and Aquatic Sciences*, 56, 1285–1292.

Lush, Gail. (2004). *Nutrition, health education and dietary reform: Gendering the 'New Science' at the Grenfell Mission*, 1893–1928. MA thesis, History, Memorial University of Newfoundland.

Lysaght, A. M. (1971). *Joseph Banks in Newfoundland and Labrador, 1766, his diary, manuscripts and collections*. Berkeley, CA.: University of California Press.

MacCall, A. D. (1985). Changes in the biomass of the California current ecosystem. In K. Sherman & L. M. Alexander (Eds.) *Variability and management of large marine ecosystems* (pp. 33–54). Boulder, CO: Westview Press.

Mahaffey, K. R. (2004). Fish and shellfish as dietary sources of methylmercury and the omega-3 fatty acids, eicosapentaenoic acid and docosahexaenoic acid: risks and benefits. *Environmental Research,* 95, 414–428.

MacNair, P. L. (1971). Descriptive notes on the Kwakiutl manufacture of eulachon oil. *Syesis* 4, 169–177.

Mann, K. H., & Lazier, J. R. N. (1996). *Dynamics of marine ecosystems: Biological–physical interactions in the oceans*. Boston, MA: Blackwell Science.

Mannion, J. J. (1977). *The peopling of Newfoundland. Essays in historical geography*. Toronto, ON: University of Toronto Press,.

Mansfield, A. W. (1967). *Seals of arctic and eastern Canada*. Bulletin No. 137 Ottawa, ON: Fisheries Research Board of Canada.

Margoshes, D. (1999). *Tommy Douglas: Building the new society*. Montreal, QC: XYZ Press.

Marshall, I. (1996). *A history and Ethnography of the Beothuk*. Montreal, QC: McGill-Queen's University Press, Montreal.

Marshall, W. H. (Ed.). (1975). *The West Coast health survey: Progress, prospects and plans.* A workbook for a multidisciplinary investigation in the province of Newfoundland. Meeting Report, September, St. John's, NL.

Martínez Murillo, Mª de las Nieves. 2003. *Size-Based Dynamics of a Demersal Fish Community: Modeling Fish–Fisheries Interactions*. PhD thesis, Memorial University of Newfoundland.

Maurer, B. A. (1999). *Untangling ecological complexity. The macroscopic perspective.* Chicago, IL: The University of Chicago Press.

McAllister, M. K., & Kirkwood, G. P. (1998). Bayesian stock assessment: a review and example application using the logistic model. *ICES Journal of Marine Science*, 55, 1031–1060.

McBride, S., & Shields, J. (1997). *Dismantling a nation: The transition to corporate rule in Canada* (*2nd ed.*) Halifax, NS: Fernwood.

McCann, L. D., (Ed.) 1982. *Heartland and hinterland: A geography of Canada.* Toronto, ON: Prentice–Hall.

McDevitt, E., Dove, M. A., Dove, R. F., & Wright, I. S. (1944). Vitamin status of the population of the West Coast of Newfoundland with emphasis on vitamin C. *Annals of International Medicine,* 20, 1–11.

McGhee, R., & Tuck, J. A. (1975). *An Archaic sequence from the Strait of Belle Isle, Labrador.* Ottawa, ON: National Museum of Canada.

McIntyre, L. (2003). Food security: More than a determinant of health. *Policy Options,* 24, 46–51.

McIntyre, L., Raine, K., & Dayle, J. B. (2001). The institutionalization of children's feeding programs in Atlantic Canada. *Canadian Journal of Dietetic Practice and Research,* 62 (2), 53–57.

McIntyre, L., Travers, K., & Dayle, J. B. (1999). Children's feeding programs in Atlantic Canada: Reducing or reproducing inequities? *Canadian Journal of Public Health,* 90, 196–200.

McMahon, T. A., & Bonner, J. T. (1983). *On size and life.* New York, NY: W.H. Freeman.

McQueen, D. J., France, R., Post, J. R., Stewart, T .J., & Lean, D. R. S. (1989). Bottom-up and top-down impacts on freshwater pelagic community structure. *Ecological Monographs*, 59, 289–309.

McRae, D. M., & Pearse, P. H. (2004). *Treaties and transition: Towards a sustainable fishery on Canada's Pacific coast.* Vancouver, BC: Hemlock Printers.

Meggs, G. (1991). *The decline of the BC Fishery.* Vancouver/Toronto: Douglas and McIntyre.

Megrey, B. A., & Wespestad, V. G. (1988). A review of biological assumptions underlying fishery assessment models. In W.S. Wooster (Ed.). *Fishery science and management: objectives and limitations.* Berlin, New York, NY: Springer-Verlag.

Melville, H. (1851). *The Whale.* New York, NY: Harper and Brothers.

Merrett, N. R., & R. L. Haedrich. (1997). *Deep-sea demersal fish and fisheries.* London: Chapman and Hall.

Metcoff, J., Goldsmith, G.A., McQueeney, A.J., Dove, R.F.,, McDevitt, E., Dove. M.A.,, & Stare, F.J.. (1945) Nutritional survey in Norris Point, Newfoundland. *Journal of Laboratory and Clinical Medicine*, 30, 475–487.

Metuzals, K. I., Wernerheim, C. M., Haedrich, R. L., Copes, P., & Murrin, A. (Submitted 2005). Data-fouling in marine fisheries: findings and a model for Newfoundland. In J. Lutz & B. Neis (Eds.), *Making and moving knowledge* (Chapter 8). Montreal, QC: McGill-Queen's University Press.

Milazzo, M. (1998). Subsidies in world fisheries: A re-examination. *World Bank Technical Paper*, 406. Washington, DC: World Bank.

Miller, J. (1984). Feasting with the Southern Tsimshian. In M. S.Anderson (Ed.), *The Tsimshian. Images of the past, views for the present.* Vancouver, BC: UBC Press.

Miller, P. J. (2003). Thailand: Profile of the world's largest farmed shrimp producer. In I. de la Torre & D. Barnhizer (Eds.), *The blues of a revolution: The damaging impacts of shrimp farming* (pp. 63–120). Tacoma, WA: The Industrial Shrimp Action Network (ISA Net) and the Asia Pacific Environmental Exchange (APEX)

Mitchell, C. L. (1982). Bioeconomics of multispecies exploitation in fisheries: management implications In M. C. Mercer (Ed.), Multispecies Approaches to Fisheries Management. *Canadian Special Publication Fisheries and Aquatic Science,* 59, 157–162.

Mitchell, H.S. (1930). A nutritional survey in Labrador and Northern Newfoundland. *Journal of the American Dietetic Association,* 6, 29–35.

Monks, G. R. (2001). Quit Blubberin': An examination of Nuu-Chah-Nulth whale butchery. *International Journal of Osteoarchaeology,* 11, 136–149.

Montevecchi, W. A. (2001). Interactions between fisheries and seabirds. In E. A. Schrieber, & J. Burger (Eds.), *The biology of marine birds (*pp. 527–557). Boca Raton, CRC Press.

Montevecchi, W. A. (2006). Influences of artificial light on marine birds. In: C. Rich and T. Longcore (Eds.), *Ecological Consequences of Artificial Night Lighting* (pp. 94-113) Washington, D.C.: Island Press.

Montevecchi, W. A., & Kirk, D. A. (1996). Great Auk. In A. Poole & F. Gill (Eds.), *Birds of North America* (No. 260). Philadelphia, PA: American Ornithologists' Union.

Montevecchi, W. A., & Myers, R. A. (1996). Dietary changes of seabirds indicate shifts in pelagic food webs. *Sarsia,* 80, 313–22.

Montevecchi, W. A., & Tuck, L. M. (1987). *Newfoundland birds: Exploitation, study, conservation.* Cambridge, MA: Nuttall Ornithological Club.

Montevecchi, W.A., & Wells, J. (2006 in press). Local names of birds in Newfoundland and Labrador. *Dialectologia et Geolinguistica.*

Morland, K., Wing, S., & Diez Roux, A. (2002). The contextual effect of the local food environment on residents' diets: The atherosclerosis risk in communities study. *The American Journal of Public Health,* 92, 1761–1767.

Morris, C., Simms, J. M., & Anderson, T. C. (2002). Biophysical overview of Gilbert Bay: a proposed marine protected area in Labrador. *Canadian Manuscript Report of Fisheries and Aquatic Sciences,* 2595, 1–25.

Morton, A., Routledge, R., Peet, C., & Ladwig, A. 2004. Sea lice (*Lepeophtheirus salmonis*) infection rates on juvenile pink (*Oncorhynchus gorbuscha*) and chum (*Oncorhynchus keta*) salmon in the nearshore marine environment of British Columbia, Canada. *Canadian Journal of Fisheries and Aquatic Science,* 61, 147–157.

Mowat, F. (1982). *Sea of slaughter.* Toronto, ON: McClellan Stewart.

Munro, G., & Sumaila, U. R. (2002). The impact of subsidies upon fisheries management and sustainability: The case of the North Atlantic. *Fish and Fisheries,* 3, 233–290.

Murawski, S. A., Maguire, J-J., Mayo, R. K., & Serchuk, F. M. (1997). Groundfish stocks and the fishing industry. In J. Boreman, B. S. Nakashima, J. A. Wilson & R. L. Kendall (Eds.), *Northwest Atlantic Groundfish: Perspectives on a Fishery Collapse.* Bethesda, MD: American Fisheries Society.

Musick, J. A., Harbin, M. M., Berkeley, S. A., Burgess, G. H., Eklund, A. M., Findley, L., et al. (2000). Marine, estuarine, and diadromous fish stocks at risk of extinction in North America (exclusive of Pacific salmonids). *Fisheries,* 25(11), 6–28.

Myers, R. A. (2001). Stock and recruitment: generalizations about maximum reproductive rate, density dependence, and variability using meta-analytic approaches. *ICES Journal of Marine Science* 58, 937–951.

Myers, R. A., Barrowman, N. J., Hutchings, J. A., & Rosenberg, A. A. (1995). Population dynamics of exploited fish stocks at low population levels. *Science*, 269, 280–283.

Myers, R. A., Hutchings, J. A., & Barrowman, N. J. (1996). Hypotheses for the decline of cod in the North Atlantic. *Marine Ecology Progress Series*, 138, 293–308.

Myers, R. A., & Worm, B. (2003). Rapid worldwide depletion of predatory fish communities. *Nature,* 423, 280–283.

NAFO, Northwest Atlantic Fisheries Organization. (1997). *Report of the Study Group on the Precautionary Approach to Fisheries Management by ICES.* NAFO SC Working Paper 97/15.

Nakashima, B. S., (1992). Patterns in Coastal Migration and Stock Structure of Capelin (*Mallotus villosus*). *Canadian Journal of Fisheries and Aquatic Science,* 49, 11, 2423–2429.

Nathoo, T., & Shoveller, J. (2003). Do healthy food baskets assess food security? *Chronic Diseases in Canada*, 24, 65–69.

National Research Council. (1999). *Sustaining marine fisheries.* Washington, DC: National Academy Press.

Naughton, P. (2005, September 28). Junk Foods to be banned in British schools. *London Times*, Online: http://www.timesonline.co.uk/article/0,,2-1547224,00.html.

Naylor, R. L., Goldburg, R. J., Primavera, J. H., Kautsky, N., Beveridge, M. C. M., Clay, J., Folke, et al.. (2000). Effect of aquaculture on world fish supplies. *Nature*, 405, 1017–1024.

Neary, Peter. (1988). *Newfoundland in the North Atlantic World, 1929–1949.* Kingston and Montreal: McGill–Queen's University Press.

Nehlsen, W., Williams, J. E., & Lichatowich, J. A. (1991). Pacific salmon at the crossroads: stocks at risk from California, Oregon, Idaho, and Washington. *Fisheries*, 16(2), 4–21.

Nelson, S. (2004). *Commercial fishing license, quota and vessel values.* Prepared for Fisheries and Oceans Canada. Vancouver, BC: Nelson Bros. Fisheries.

New York Nutrition Council. (1921). Report of Committee on Training Standards. *Journal of Home Economics,* 13.10, 493

Newell, D., & Ommer, R. E. (Eds.) (1999). *Fishing places, fishing people: Traditions and issues in Canadian small-scale fisheries.* Toronto, ON: University of Toronto Press.

Newfoundland, Department of Health. (n.d.). *Some facts on public health nutrition in Newfoundland.* [Typescript] 7pp.

Newfoundland and Labrador Community Accounts. (n.d.). Community Accounts [online data]. *http://www.communityaccounts.ca/communityaccounts/onlinedata/default.asp.*

Newfoundland and Labrador, Government of. (1986). *Building on our strengths*: *the final report of the Royal Commission on employment and unemployment.* St. John's, NL: Queen's Printer.

Newfoundland Nutrition Council. (1944). *Provincial Archives of Newfoundland and Labrador.* GN 38 56-1-8 file 4.

Newfoundland Nutrition Council. (1945). Minutes. Provincial Archives of Newfoundland and Labrador. GN 38 56-1-8 file 26.

Newfoundland, Public Health and Welfare (1935). *"Population by Districts and Settlements."* Tenth Census of Newfoundland and Labrador, 1935 (vol.1) St John's: Evening Telegram Ltd. 1937, 168–169.

NFA, National Food Alliance. 1997. *Myths about food and low income.* London: National Food Alliance Food Poverty Project.

Norris Point Study. (1945). Provincial Archives of Newfoundland and Labrador. GN 38 56-1-8-file 3.

North Sea Task Force. (1993). *North Sea quality status report 1993.* Oslo and Paris Commissions. Fredensborg, Denmark London: Olsen and Olsen, 174 pp.

Nutrition Reviews. (1945). Nutrition in Newfoundland. *Nutrition Reviews,* 3, 8 August, 251–53.

Nutritional Survey of East Coast Newfoundland. (1944). Provincial Archives of Newfoundland and Labrador. GN 38 56-1-9 file 1.

Nuxalk Food and Nutrition Program. (1984). *Nuxalk food and nutrition handbook.* Bella Coola, BC: Nuxalk Nation.

O'Connor, J. S. (1998). Social justice, social citizenship, and the welfare state: Canada in comparative context. In R. Helmes-Hayes, & J. Curtis (Eds.), *The vertical mosaic revisited* (pp. 180–231). Toronto: University of Toronto Press.

Odum, E. P. (1985). Trends expected in stressed ecosystems. *BioScience*, 35, 419–422.

Olds, J. (1943). Vitamin C. *Northern Medical Review,* 1/1, 1–6.

Olsen, S. B. (2003). Frameworks and indicators for assessing progress in integrated coastal management initiatives. *Ocean and Coastal Management,* 46, 347–361.

Olsen, Y. (1999). Lipids and essential fatty acids in aquatic food webs: What can freshwater ecologists learn from mariculture? In M. T. Arts & B. C. Wainman (Eds.), *Lipids in Freshwater Ecosystems* (pp. 161–202). New York: Springer-Verlag.

Olver, C. H., Shuter, B. J., & Minns, C. K. (1995). Toward a definition of conservation principles for fisheries management. *Canadian Journal of Fisheries and Aquatic Sciences,* 52, 1584–1594.

Ommer, R. E. (2002a). Newfoundland: Environment, history, and rural development. In: Ommer, R.E. (Ed.), *The resilient outport: ecology, economy, and society in rural Newfoundland* (pp. 21–39). St. John's, NL: ISER Books.

Ommer, R. E. (Ed.) (2002b). *The resilient outport.* St. John's, NL: ISER Books.

Ommer, R. E. (2000). The ethical implications of property concepts in a fishery. In H. Coward, R. E. Ommer & T. J. Pitcher (Eds.), *Just fish: Ethics and Canadian marine fisheries* (pp. 117–139). St. John's, NL: ISER Books.

Ommer, R. E. (1994). One hundred years of fishery crisis in Newfoundland. *Acadensis,* 23, 5–20.

Ommer, R. E. (1991). *From outpost to outport. A structural analysis of the Jersey–Gaspé Cod Fishery, 1767–1886.* Kingston, ON: McGill-Queens University Press.

Ommer, R. E. (1990a). Merchant credit and the informal economy: Newfoundland, 1919–1929. Historical Papers, Quebec 1989, Canadian Historical Association, 167–89.

Ommer, R. E. (Ed.) (1990b). *Merchant credit and labour strategies in historical perspective.* Fredericton, NB: Acadiensis Press.

Ommer, R. E., & the *Coasts Under Stress* Research Project Team. (In press). *Coasts Under Stress: Restructuring and social-ecological health.* Montreal: McGill-Queens Press.

Ommer, R. E., & Sinclair, P. R. 1999. Outports under threat: social roots of systemic crisis in rural Newfoundland. In Reginald Byron & John Hutson (Eds.), *Local enterprise on the North Atlantic margin* (pp. 253–275). Aldershot: Ashgate.

Ommer, R. E., & Turner, N. J. (2004). Informal rural economies in history. *Labour/Le Travail,* 53, 127–157.

Omohundro, J. T. (1999). Living off the land. In L. F. Felt & P. R. Sinclair (Eds.), *Living on the edge: The great northern peninsula of Newfoundland,* (pp. 103–127). St. John's, NL: ISER Books.

Omohundro, J. T. (1994). *Rough food. The seasons of subsistence in Northern Newfoundland.* St. John's, NL: ISER Books.

O'Neil, J. (1986). Colonial stress in the Canadian Arctic: An ethnography of young adults changing. In C. Janes, R. Stall, & S. Gifford (Eds.), *Anthropology and Epidemiology* (pp. 249–274). Dordrecht, Holland: D. Reidel Publishers.

Orensanz, J. M. L., Armstrong, J., Armstrong, D., & Hilborn, R. (1998). Crustacean resources are vulnerable to serial depletion—the multifaceted decline of crab and shrimp fisheries in the Greater Gulf of Alaska. *Reviews in Fish Biology and Fisheries,* 8, 117–176.

Orth, R. J., & Moore, K. A. (1983). Chesapeake Bay: an unprecedented decline in submerged aquatic vegetation. *Science,* 222, 51–53.

Osborne, B. S. (2001). Landscapes, memory, monuments and commemoration: Putting identity in its place. *Canadian Ethnic Studies,* 33, 39–77.

O'Sullivan, M., & O'Morain, C. A. (1998). Nutritional therapy in Crohn's disease. *Inflammatory Bowel Disease,* 4, 45–53.

Overton, J. (1998). Brown flour and beriberi: The Politics of dietary and health reform in Newfoundland in the first half of the Twentieth Century. *Newfoundland Studies*, 14, 1, 14–20.

Paine, R. T. (1994). *Marine rocky shores and community ecology: An experimentalist's perspective.* Oldendorf/Luhe, Germany: Ecology Institute.

Paine, R. T. (1984). Some approaches to modelling multispecies systems. In R. M. May (Ed.), *Exploitation of Marine Communities* (pp. 191–207). Berlin: Springer-Verlag.

Parrish, C. C., Abrajano, T. A., Budge, S. M., Helleur, R. J., Hudson, E. D., Pulchan, K., & Ramos, C. (2000). Lipid and phenolic biomarkers in marine ecosystems: analysis and applications. In P. J. Wangersky (Ed.), *Marine Chemistry* (pp. 193–223). Heidelberg: Springer-Verlag,

Parrish, C. C., Pathy, D. A., & Angel, A. (1990). Dietary fish oils limit adipose tissue hypertrophy in rats. *Metabolism,* 39, 217–219.

Parrish, C. C., Zsigmond, E. M., Pathy, D. A. , Shaikh, N. A., Fong, B. S., & Angel, A. (1989). Effect of fish oil diets on lipid molecular structure of adipocyte plasma membranes in rats. In R. K. Chandra (Ed.), *Health Effects of Fish and Fish Oils* (pp. 159–169). St. John's, NL: ARTS Biomedical Publishers and Distributors.

Parsons, T. R. (1992). The removal of marine predators by fisheries and the impact of trophic structure. *Marine Pollution Bulletin,* 25, 51–53.

Pauly, D. (1995). Anecdotes and the shifting baseline syndrome of fisheries. *Trends in Ecology and Evolution*, 10, 430.

Pauly, D. (1988). Fisheries research and the demersal fisheries of Southeast Asia. In J. A. Gulland (Ed.), *Fish Population Dynamics* (pp. 329–348). London, UK: Wiley and Sons.

Pauly, D., Christensen, V., Dalsgaard, J., Froese, R., & Torres, F. Jr. (1998). Fishing down marine food webs. *Science,* 279, 860–863.

Pauly, D., & Palomares M. L. (2001). Fishing down marine food webs: an update. In L. Bendell-Young & P. Gallaugher (Eds.), *Waters in peril* (pp. 47–56). Norwell, MA: Kluwer Academic Publishers.

Pauly, D, Palomares, M. L., Froese, F., Sa-a, P., Vakily, M., Preikshot, D., & Wallace, S. (2001). Fishing down Canadian aquatic food webs. *Canadian Journal of Fisheries and Aquatic Science*, 58(1), 51–62.

Pauly, D., Watson, R., & Alder, J. (2005). Global trends in world fisheries: impacts on marine ecosystems and food security. *Philosophical Transactions of the Royal Society of London, Series B*, 360, 5–12.

Peavy, V. (1992). New concepts and practices in career counselling: A research and development project. *National Consultation on Vocational Counselling, 18*. Toronto, ON: OISE Press.

Pendry, S. (1998). Sustainability—An elusive and misused concept—reply. *ORYX*, 32 (2), 88.

Penhale, P. A., & Thayer, G.W. (1980). Uptake and transfer of carbon and phosphorus by eelgrass (Zostera marina) and its epiphytes. *Journal of Experimental Marine Biology and Ecology,* 42, 113–123.

Perry, A. L., Low, P. J., Ellis, J. R., & Reynolds, J. D. (2005). Climate change and distribution shifts in marine fishes. *Science,* 308, 1912–1915.

Peters, H. S., & Burleigh, T. D. (1951). *The birds of Newfoundland.* St. John's, NL: Department of Natural Resources.

Peters, R. H. (1983). *The ecological implications of body size.* New York: Cambridge University Press.

Petrini, C. (2005). Buono, Pulito, Egiusto ["Good, Clean, Fair"]. Princip di nuova gastronomia. Slow Food International, Bra, Italy.

Phillis, Y. (2001). Sustainability, an ill-defined concept and its assessment. *Ecological Economics*, 37, 435–456.

Piatt, J. F., & Nettleship, D. N. (1987). Incidental catch of marine birds and mammals in fishing nets off Newfoundland, Canada. *Marine Pollution Bulletin,* 18, 344–349.

Piatt, J. F., Nettleship, D. N., & Threlfall, W. (1984). Net mortality of Common Murres and Atlantic Puffins in Newfoundland, 1951–1981. In D .N. Nettleship, G. A. Sanger & P.F. Springer (Eds.), *Marine birds: Their feeding ecology and commercial fishery relationships* (pp. 196–206). Ottawa, ON: Canadian Wildlife Service Special Publication.

Pikitch, E. K., Santora, C., Babcock, E.A. Bakun, A., Bonfil, R., Conover, D.O., et al. (2004). Ecosystem-based fishery management, *Science,* 305, 346–347.

Pimm, S. L., & Hyman, J. B. (1987). Ecological stability in the context of multispecies fisheries. *Canadian Journal of Fisheries and Aquatic Science*, 44 (Suppl.2), 84–94.

Pitcher, T. J. (2005). "Back to the Future": A fresh policy initiative for fisheries and a restoration ecology for ocean ecosystems. *Philosophical Transactions of the Royal Society. Series B. Biological Sciences*, 360(1453), 107–121.

Pitcher, T. J. (2004a). *Introduction to the methodological challenges in 'Back-To-The-Future' research.* In T. J. Pitcher (Ed.), Back to the Future: Advances in methodology for modelling and evaluating past ecosystems as future policy goals (pp. 4–10). UBC Fisheries Centre Research Reports, 12(1).

Pitcher, T. J. (2004b). *Why we have to open the lost valley: criteria and simulations for sustainable fisheries.* In T. J. Pitcher (Ed.), Back to the Future: Advances in methodology for modelling and evaluating past ecosystems as future policy goals (pp. 78–86). UBC Fisheries Centre Research Reports, 12(1).

Pitcher, T. J. (2004c). *The problem of extinctions.* In T. J. Pitcher (Ed Back to the Future: Advances in methodology for modelling and evaluating past ecosystems as future policy goals (pp. 21–28). UBC Fisheries Centre Research Reports, 12(1).

Pitcher, T. J. (2001). Fisheries managed to rebuild ecosystems: Reconstructing the past to salvage the future. *Ecological Applications*, 11, 601–617.

Pitcher, T. J. (2000). Fisheries management that aims to rebuild resources can help resolve disputes, reinvigorate fisheries science and encourage public support. *Fish and Fisheries*, 1, 99–103.

Pitcher, T. J. (1998a). A cover story: Fisheries may drive stocks to extinction. *Reviews in Fish Biology and Fisheries*, 8, 367–370.

Pitcher, T. J. (1998b). 'Back to the Future': A novel methodology and policy goal in fisheries.. In D. Pauly, T. J. Pitcher, & D. Preikshot (Eds.), *Back to the Future: Reconstructing the Strait of Georgia ecosystem* (pp. 4–7). UBC Fisheries Centre Research Reports, 6(5).

Pitcher, T. J., & Ainsworth, C. (In press). Back-to-the-Future: a candidate ecosystem-based solution to the fisheries problem. In Nielson J. (Ed.), *Reconciling fisheries with conservation: Proceedings of the 4th World Fisheries Congress*. Bethesda, USA: American Fisheries Society.

Pitcher, T. J., Ainsworth, C., Buchary, E. A. Cheung, W. W. L., Forrest, R. Haggan, N., Lozano, H., Morato, T., & Morissette, L. (2005). Strategic management of marine ecosystems using whole-ecosystem simulation modelling: The "Back to the Future" policy approach. In E. Levner, I. Linkov, & J. M. Proth (Eds.), *Strategic management of marine ecosystems* (pp. 199–258). NATO Science Series IV. Earth and Environmental Sciences, 50.

Pitcher, T. J., & Alheit. J. (1995). What makes a hake? A review of the critical biological features that sustain global hake fisheries. In T. J. Pitcher, & J. Alheit (Eds.), *Hake: fisheries, ecology and markets* (pp. 1–15). London: Chapman and Hall.

Pitcher, T. J., Power, M., & Wood, L. (Eds.) (2002a). *Restoring the past to salvage the future: report on a community participation workshop in Prince Rupert*, BC. UBC Fisheries Centre Research Reports, 10(7).

Pitcher, T. J., Buchary, E. A., & Sumaila, U. R. (2002b). *A synopsis of Canadian fisheries.* UBC Fisheries Centre. *http://www.fisheries.ubc.ca/publications/reports/canada-syn.pdf*

Pitcher, T. J., Heymans, J. J., Brignall, C., Vasconcellos, M., & Haggan, N. (2002c). Information supporting past and present ecosystem models of the Hecate Strait and the Newfoundland shelf. Fisheries Centre Research Reports 10 (1) 116pp.

Pitcher, T. J., Heymans, J. J., Ainsworth, C., Buchary, E. A., Sumaila, U. R., & Christensen, V. (2002d). Opening the lost valley: Implementing A 'Back To Future' restoration policy for marine ecosystems and their fisheries. In E. E. Knudsen, D. D. MacDonald, & J. K. Muirhead (Eds.), *Fish in the Future? Perspectives on fisheries sustainability*. Bethesda, MD: American Fisheries Society.

Pitcher, T. J., Buchary, E., & Trujillo, P. (Eds.) (2002e). *Spatial simulations of Hong Kong's marine ecosystem: Ecological and economic forecasting of marine protected areas with human-made reefs*. UBC Fisheries Centre Research Reports, 10(3).

Pollack, S. L. (2001). Consumer demand for fruit and vegetables: The U.S. example. In A. Regmi, (Ed.), *Changing structure of food consumption and trade* (pp. 49–54). Market and Trade Economics Division, Economic Research Service, U.S. department of Agriculture, Agriculture and Trade Report, WRS-01-1.

Pomeroy, L. R., Wiebe, W. J., Deibel, D., Thompson, R. J., Rowe, G. T., & Pakulski, J. D. (1991). Bacterial responses to temperature and substrate concentration during the Newfoundland spring bloom. *Marine Ecology Progress Series,* 75, 143–159.

Post, J. R., Sullivan, M., Cox, S., Lester, N. P., Walters, C. J., Parkinson, E. A., Paul, A. J., Lackson, L., & Shuter, B. J. (2002). Canada's recreational fisheries: the invisible collapse. *Fisheries*, 27 (1), 6–19.

Powell, T. M. (1989). Physical and biological scales of variability in lakes, estuaries, and the coastal ocean. In J. Roughgarden, R. M. May, & S. A. Levin (Eds.), *Perspectives in ecological theory* (pp. 157–176). Princeton, NJ: Princeton Univ. Press.

Prowse, D. W. (2002 orig. 1895). *A history of Newfoundland*. Portugal Cove—St. Philip's, NL: Boulder Publications.

Puccia, C. T., & Levins, R. (1985). *Qualitative modelling of complex systems: An introduction to loop analysis and time averaging*. Cambridge, MA.: Harvard University Press.

Raffoul, F. (2002). Heidegger and the origins of responsibility. In F. Raffoul, & D. Pettigrew (Eds.), *Heidegger and practical philosophy* (pp. 205–218). New York: SUNY Press.

Ralph, A. (1998). Unequal health. In S. Griffiths & J. Wallace (Eds.), *Consuming patterns: Food in the age of anxiety* (pp. 88–96). Manchester, UK: Mandolin.

Rapport, D. J., Regier, H. A., & Hutchinson, T. C. (1985). Ecosystem behaviour under stress. *American Naturalist*, 125, 617–640.

Rapport, D. J., & Whitford, W.G. (1999). How ecosystems respond to stress. *BioScience,* 49 , 193–203.

Reeves, R. R. (2002). The origins and character of 'aboriginal subsistence' whaling: a global review. *Mammal Review*, 32, 71–106.

Regier, H. A. (1973). Sequence of exploitation of stocks in multispecies fisheries in the Laurentian Great Lakes. *Journal of the Fisheries Research Board of Canada,* 30, 1992–1999.

Regier, H. A., & Loftus, K. H. (1972). Effects of fisheries exploitation on salmonid communities in oligotrophic lakes. *Journal of the Fisheries Research Board of Canada*, 29, 959–968.

Regmi, A. (Ed.). (2001). *Changing structure of food consumption and trade*. Market and Trade Economics Division, Economic Research Service, U.S. department of Agriculture, Agriculture and Trade Report, WRS-01-1.

Regmi, A., & Dyck, J. (2001). Effects of urbanization on global food demand. In A. Regmi (Ed.), *Changing structure of food consumption and trade* (pp. 23–30). Market and Trade Economics Division, Economic Research Service, U.S. department of Agriculture, Agriculture and Trade Report, WRS-01-1.

Reuss, N., & Poulsen, L. K. (2002). Evaluation of fatty acids as biomarkers for a natural plankton community. A field study of a spring bloom and a post-bloom period off West Greenland. *Marine Biology,* 141, 423–434.

Reuters. (1999). Argentina Hake Ban and Protest http://segate.sunet.se/cgi-bin/wa?A2=ind9906andL=fish-sciandF=andS=andP=5056

Rice, J. C. (2000). Evaluating fishery impacts using metrics of community structure. *ICES Journal of Marine Science,* 57, 682–688.

Richardson, M., & Green, B. (1989). The Fisheries Co-Management Initiative in Haida Gwaii. In E. Pinkerton (Ed.), *Cooperative management of local fisheries: New directions for improved management and community development* (pp. 249–261). Vancouver, BC: UBC Press.

Riches, G. (2002). The human right to food. In D. Lamberton (Ed.), *Managing the global: Globalization, employment and quality of life* (pp. 169–175). London: I.B. Tauris.

Riches, G. (2000). *Hunger, welfare and food security: Some possibilities for social work.* Ottawa, ON: Love Printing.

Riches, G. (1986). The collapse of the public safety net. In F. Riches (Ed.), *Food banks and the welfare crisis* (pp. 59–112). Ottawa, ON: Love Printing.

Ritz, D. A., Lewis, M. E., & Shen, M. (1989). Response to organic enrichment of infaunal macrobenthic communities under salmonid sea cages. *Marine Biology,* 103, 211–214.

Robertson, G. J., Wilhelm, S. I., & Taylor, P. A. (2004). Population size and trends of seabirds breeding on Gull and Great Islands, Witless Bay Islands Ecological Reserve, Newfoundland, up to 2003. Canadian Wildlife Service Technical Report Series No. 418. Atlantic Region.

Robinson, J., & Bennett, E. (Eds.) (1999). *Hunting for Sustainability in Tropical Forests.* New York: Columbia University Press.

Rockne, J. (2002). *Branded: corporations and our schools.* http://www.reclaimdemocracy. org/education/branded_schools.html (accessed 2005-12-13)

Roebothan, B. V. (2003). *Nutrition Newfoundland and Labrador: The report of a survey of residents of Newfoundland and Labrador, 1996.* St. John's, NL: Department of Health and Community Services, Province of Newfoundland and Labrador.

Rojek, C. (1995). *Decentring leisure: Rethinking leisure theory.* London: Sage.

Roman, J., & Palumbi, S. R. (2003). Whales before whaling in the North Atlantic. *Science,* 301, 508–510.

Rompkey, R. (1991). *Grenfell of Labrador: A biography.* Toronto, ON: University of Toronto Press.

Rosenberg, A., Mooney-Seuss, M., & Ninnes, C. (2005). *Bycatch on the High Seas: A Review of the effectiveness of the Northwest Atlantic Fisheries Organization. Prepared for World Wildlife Fund Canada.* Tampa, FL: MTAG Americas, Ltd.

Roughgarden, J. (1998). *Primer of ecological theory.* Upper Saddle River, NJ: Prentice Hall.

Roy, N. (1998). The Newfoundland fishery: A descriptive analysis. *Marine Resource Economics,* 13, 197–213.

Ruzzante, D. E., Wroblewski, J. S., Taggart, C. T., Smedbol, R. K., Cook, D., & Goddard, S. V. (2000). Bay-scale population structure in coastal Atlantic cod in Labrador and Newfoundland, Canada. *Journal of Fish Biology,* 56, 431–447.

Safina, C. (1995). The world's imperilled fish. *Scientific American,* 273, 46–53.

Saunders, G. L. (1994). *Dr. Olds of Twillingate: Portrait of an American surgeon in Newfoundland.* St. John's, NL: Breakwater.

Schindler, D. E., T. E. Essington, J. F. Kitchell, C. Boggs, and R. Hilborn. (2002). Sharks and tunas: fisheries impacts on predators with contrasting life histories. *Ecological Applications*, 12, 735–748.

Schluter, G., & Lee, C. (1999). Changing food consumption patterns: Their effect on the U.S. food system, 1972–92. *Food Review,* (January–April), 35–37.

Schmidhuber, J. (2003). *The outlook for long-term changes in food consumption patterns: Concerns and policy options.* Paper prepared for the FAO Scientific Workshop on Globalization of the Food System: Impacts on Food Security and Nutrition, FAO, Rome, 8–10 October. Retrived from www.fao.org/es/ESD/schmidhuberdiets.pdf 09/01/05.

Schmidt-Nielsen, K. (1984). *Scaling. Why is animal size so important?* New York: Cambridge University Press. Cambridge.

Scholliers, P. (Ed.). (2001). *Food, drink and identity: Cooking, eating and drinking in Europe since the Middle Ages.* New York and Oxford: Berg.

Schrank, W., Skoda, B., Roy, N., & Tsoa, E. (1987). Canadian government financial involvement in a marine fishery: the case of Newfoundland, 1972/73–1980–81. *Ocean Development and International Law*, 18, 533–584.

Scobie, K, Burke, B. S., & Stuart, H. C. (1949) Studies of Nutrition in Newfoundland Children. *Canadian Medical Association Journal*, 60, 233–234.

SCOPE, Scientific Committee on Problems of the Environment. 1999. *Report of the Workshop to Assess the Role of Soil and Sediment Biodiversity in the Functioning of Critical Transition Zones.* 27 August– 2 September, Corvallis, OR. *http://www.nrel.colostate.edu/projects/soil/SCOPE/scope99summary.htm*

Severs, D., Williams, T., & Davies, J. W. (1961). Infantile Scurvy—A Public Health Problem. *Canadian Journal of Public Health,* 52, 214–320.

Shahidi, F., & Wanasundara, U. (1998). Seal oil as a novel nutraceutical. Abstract 27. St. John's, NL: 43rd Atlantic Fisheries Technological Conference, 1998.

Shepherd, C. J., Pike, I. H., & Barlow, S. M. (2005). *Sustainable feed resources of marine origin.* European Aquaculture Society. Special Pub. No. 35, 59–66. Belgium.

Sherman, K. (1994). Sustainability, biomass yields, and health of coastal ecosystems: an ecological perspective. *Marine Ecology Progress Series*, 112, 277–301.

Short, F. T., & Wyllie-Echeverria, S. (1996). Natural and human-induced disturbance of seagrasses. *Environmental Conservation,* 23, 17–27.

Sinclair, P. R. (1985). *From Traps to Draggers: Domestic Commodity Production in Northwest Newfoundland, 1850–1982.* St. John's, NL: ISER Books.

Siscovick, D. S., Raghunathan, T. E., King, I., Weinmann, S., Bovbjerg, V. E., Kushi, L., *et al.* (2000). Dietary intake of long-chain n-3 polyunsaturated fatty acids and the risk of primary cardiac arrest. *American Journal of Clinical Nutrition*, 71, 208–212.

Silvert, W. (1992). Assessing environmental impacts of finfish aquaculture in marine waters. *Aquaculture*, 107, 67–79.

Smith, T. D. (1994). *Scaling fisheries.* Cambridge: Cambridge Univ. Press.

Smith, P. E. I. (1987). In winter quarters. *Newfoundland Studies*, 3, 1–36.

Smith, D., & Nicolson, M. (1997). Nutrition, education, ignorance and Income: A Twentieth Century debate. In H. Kamminga, & A. Cunningham (Eds.), *The science and culture of nutrition, 1840s–1940s.* Amsterdam: Rodopi.

Smith, R. L., & Smith, T. M. (1998). *Elements of ecology*. San Francisco, CA:Addison Wesley Longman. Inc.

Smith, C. R., Austen, M. C, Boucher, G., Heip C., Hutchings, P. A., King, G. M., Koike, I., Lambshead, P.J.D., & Snelgrove, P. 2000. Global change and biodiversity linkages across the sediment–water interface. *BioScience* 50, 1108–1120.

Smyth, J. (Ed.) (1993). *A socially critical view of the self-managing school*. New York, NY: Falmer.

Sniffen, S. B. (1923). The travelling Labrador health unit. *Among the Deep Sea Fishers*, Jan. 1923, 109–114.

Soto, H. (2004). "Colapso pesquero" Adiós a la merluza y al jurel. *Punto Final*, Chile, 569, 11–24 June. http://www.puntofinal.cl/569/merluzayjurel.htm.

Southard, F. E. (1982). *Salt cod and God: An ethnography of socio-economic conditions affecting status in a southern Labrador community*. MA thesis, Anthropology, Memorial University of Newfoundland.

Springer, A. M., Estes, J. A., van Vliet, G. B., Williams, T. M., Doak, D. F., Danner, E. M., Forney, K. A., & Pfister, B. (2003). Sequential megafaunal collapse in the North Pacific: an ongoing legacy of industrial whaling? *Proceeding National Academy of Science USA*, 100, 12223–12228.

Statistics Canada. (2004). *Profile of census divisions and subdivisions—2001 census of Canada*. Ottawa, ON: Industry Canada. Catalogue no. 95-21S-XPB.

Steele D. H., Green, J., & Anderson, R. (1992). The managed commercial annihilation of Northern Cod. *Newfoundland Studies,* 8, 34–68.

Steele, J. H. (1984). Kinds of variability and uncertainty affecting fisheries. In R. M. May (ed.) *Exploitation of Marine Communities* (pp. 245–262). Berlin: Springer-Verlag.

Steele, J. H. (1978). Some comments on plankton patches. In J. H. Steele (Ed.), *Spatial Pattern in Plankton Communities* (pp. 1–20). New York: Plenum Press.

Stenhouse I. J., Burke, C., Davoren G., & Montevecchi, W. A. (2002). *What the past can tell us about the future: Part II Historic reconstruction of the seabird community of the Newfoundland Shelf, 1500-2000*. Poster Presentation Ocean Management Research Network (OMRN) Conference, Ottawa.

Stephenson, P., Hopkinson, J., & Turner, N. J. (1995). Changing traditional diet and nutrition of Aboriginal peoples of Coastal British Columbia., In P. H. Stephenson, S. J. Elliott, L. T. Foster, & J. Harris (Eds.), *A Persistent spirit: Towards understanding Aboriginal health in British Columbia* (pp. 129–165). Victoria, BC: University of Victoria, Western Geographic Series No. 31.

Stephenson, R. L., & Lane, D. E. (1995). Fisheries management science: a plea for conceptual change. *Canadian Journal of Fisheries and Aquatic Sciences*, 52, 2051–2056.

Stewart, J. E. (1997). Environmental impacts of aquaculture. *World Aquaculture*, 28, 47–52.

Strategic Communications Inc. (2001). BC-wide marine protected areas poll. October 10–13, 2001. Online at: *http://www.cpawsbc.org/press/20011122.php*.

Sugihara, G., Garcia, S., Guilland, J. A., Lawton, J. H., Maske, H., Paine, R. T., *et al.* (1984). Ecosystem dynamics. In R. M. May (ed.) *Exploitation of marine communities* (pp. 130–153). Berlin: Springer-Verlag.

Sumaila, U. R. (2004). Intergenerational cost benefit analysis and marine ecosystem restoration. *Fish and Fisheries*, 5, 329–343.

Sumaila, U. R., Alder, J., Ishimura, G., Cheung, W. W. L., Dropkin, L., Hopkins, S., Sullivan, S., & Kitchingman, A. (In press). *Values from marine ecosystems of the United States*. UBC Fisheries Centre Research Reports.

Sumaila, U. R., & Bawumia, M. (2000). Ecosystem justice and the marketplace. In H. Coward, R. E. Ommer & T. J. Pitcher (Eds.), *Just Fish: Ethics and Canadian Marine Fisheries* (pp. 140–153). St. Johns, NL: ISER Books.

Sumaila, U. R., & Walters, C. (2005). Intergenerational discounting: A new intuitive approach. *Ecological Economics*, 52, 135–142.

Sun, M. Y., Aller, R. C., Lee, C., & Wakeham, S. G. (1999). Enhanced degradation of algal lipids by benthic macrofaunal activity: Effect of Yoldia limatula. *Journal of Marine Research,* 57, 775–804.

Suttles, W. (1951). The early diffusion of the potato among the Coast Salish. *Southwestern Journal of Anthropology,* 7, 272–288.

Svirezhev, Y. M. 1998. Sustainable biosphere—critical overview of basic concepts of sustainability. *Ecological Modelling,* 106 (1), 47–61.

Sweeney, W. A. (2003). *Happy and healthy in a chemical world. Learn about food, the environment and your safety*. Larkspur, CA: 1st Books Library.

Sy, A., Rhein, M., Lazier, J. R. N., Koltermann, K. P.; Meincke, J., Putzka, A., & Bersch, M. (1997). Surprisingly rapid spreading of newly formed intermediate waters across the North Atlantic Ocean, *Nature*, 386, 675–79.

Tanner, V. (1944). *Outlines of the geography, life and customs of Newfoundland-Labrador*. Helsinki: Acta Geographica 8, No. 1.

Taylor, S., Fazal, R., Lingard, B., & Henry, M. (1997). Globalisation, the state and education policy making. In *Educational policy and the politics of change* (pp. 54–77). London: Routledge.

Templeman, W. (1966). Marine resources of Newfoundland. Ottawa: Fisheries Research Board of Canada, Bull. 154.

Thom, R. M., Antrim, L. C., Borde, A. B., Gariner, W. W., Shreffler, D. K., Farley, P. G., Norris, J. G., Wyllie-Echeverria, S., & McKenzie, T. P. (1998). Puget Sound's eelgrass meadows: Factors contributed to depth distribution and spatial patchiness. *Puget Sound Research '98, Conference Proceedings*, pp. 363–370.

Thomas, M. (2005). Are universities hooked on Coca Cola? *The Ubyssey*, 1 February. *http://www.ubyssey.bc.ca/20050201.*

Thornton, P. A. 1979. *Dynamic equilibrium: Population, ecology and economy in the Strait of Belle Isle,* Newfoundland, 1840–1940. Ph.D. dissertation, Aberdeen University, Scotland.

Tisdale, M. J. (1992). Essential fatty acids and cancer. In A. Sinclair & R. Gibson (Eds.), *Essential fatty acids and eicosanoids* (pp. 389–392). Champaign, IL: American Oil Chemists' Society.

Townson, M. (1999). *Health and wealth: How social and economic factors affect our well-being*. Toronto: Lorimer.

Tsutsumi, H., Kikuchi, T., Tanaka, M., Higashi, T., Imasaka, K., & Miyazaki, M. (1991). Benthic faunal succession in a cove organically polluted by fish farming. *Marine Pollution Bulletin,* 23, 233–238.

Tuck, J. A. (1976). *Ancient people of Port au Choix*. Toronto, ON: Van Nostrand Reinhold.

Turner, N. J. (2005). *The earth's blanket. Traditional teachings for sustainable living.* Vancouver, BC: Douglas and McIntyre and Seattle: University of Washington Press.

Turner, N. J. (2004). *Plants of Haida Gwaii.* Xaadaa Gwaay guud gina ḵ'aws (Skidegate), X̲aadaa Gwaayee guu giin ḵ'aws (Massett). Winlaw, BC: Sono Nis Press.

Turner, N. J. (2003a). "Passing on the News": Women's work, traditional knowledge and plant resource management in indigenous societies of NW N. America. In P. Howard (Ed.), *Women and plants: Case studies on gender relations in local plant genetic resource management* (pp. 133–149). UK: Zed Books.

Turner, N. J. (2003b). The Ethnobotany of "edible seaweed" (*Porphyra abbottiae* Krishnamurthy and related species; Rhodophyta: Bangiales) and its use by First Nations on the Pacific Coast of Canada. *Canadian Journal of Botany,* 81, 283–293.

Turner, N. J. (1995). *Food plants of Coastal First Peoples.* Victoria, BC: Royal British Columbia Museum and Vancouver: UBC Press.

Turner, N. J. (1978). Plants of the Nootka Sound Indians as recorded by Captain Cook. *Sound Heritage,* 7, 78–87.

Turner, N. J., & Berkes, F. (2006). Coming to understanding: Developing conservation through incremental learning. *Human Ecology,* 34(4), in press.

Turner, N. J., & Clifton, H. (In press). "The forest and the seaweed": Gitga'at seaweed, traditional ecological knowledge and community survival. In C. Menzies (Ed.), *Integrating local level ecological knowledge with natural resource management: Exploring the possibilities and the obstacles.* Lincoln, NE: University of Nebraska.

Turner, N. J., & Davis, A. (1993). "When everything was scarce": The role of plants as famine foods in northwestern North America. *Journal of Ethnobiology,* 13(2), 1–28.

Turner, N. J., Ignace, M. B., & Ignace, R. (2000). Traditional ecological knowledge and wisdom of Aboriginal peoples in British Columbia. In J. Ford & D. R. Martinez (Eds.), *Ecological Applications,* 10 (5), 1275–1287.

Turner, N. J., & Ommer, R. E. (2004). "Our Food is our medicine": Traditional plant foods, traditional ecological knowledge and health in a changing environment. In A. Wong (Ed.), *First Nations Nutrition and Health Conference Proceedings* (pp. 22–39). Conference June 2003, Squamish Nation, Vancouver, BC. Vancouver, BC: Arbokem Inc.

Turner, N. J., & Szczawinski, A. F. (1991). *Common poisonous plants and mushrooms of North America.* Portland, OR: Timber Press.

Turner, N. J., Thomas, J., Carlson, B. F., & Ogilvie, R. T. (1983). *Ethnobotany of the Nitinaht Indians of Vancouver Island.* Victoria: British Columbia Provincial Museum Occasional Paper No. 24.

Turner, N. J., & Wilson, B. K. (2003). *K'aaw k'iihl*: A time-honoured tradition for today's world. In S. Foster & H. Arntzen (Eds.), *Handbook on recycling.* Victoria, BC: Artists' Response Team.

Ueno, D., Kajiwara, N., Tanaka, H., Subramanian, A., Fillman, G., Lam, P. K. S. *et al.* (2004). Global pollution monitoring of polybrominated diphenyl ethers using skipjack tuna as a bioindicator. *Environmental Science and Technology,* 38, 2312–2316.

Ulltang, Ø. (1996). Stock assessment and biological knowledge: can prediction uncertainty be reduced? *ICES Journal of Marine Science,* 53, 659–675.

Umegaki, K., Ikegami, S., & Ichikawa, T. (1995). Fish oil enhances pentachlorobenzene metabolism and reduces its accumulation in rats. *Journal of Nutrition,* 125, 147–153.

van Biesen, G., & Parrish, C. C. (2005). Long-chain monounsaturated fatty acids as biomarkers for the dispersal of organic waste from a fish farm. *Marine Environmental Research*, 60, 375–388.

Vanni, M. J., Luecke, C., Kitchell, J. F., Allen, Y., Temte, J., & Magnuson, J. J. (1990). Effects on lower trophic levels of massive fish mortality. *Nature,* 344, 333–335.

van Oostdam, J., Gilman, A., Dewailly, E., Usher, P., Wheatley, B. Kuhnlein, H. V., Neve, S., Walker, J., Tracy, B., Feeley, M., Jerome, V., & Kwavnick, B. (1999). Human health implications of environmental contaminants in Arctic Canada: a review. *Science of the Total Environment,* 230, 1–82.

Vaughn, M., & Mitchell, H. S. (1933). A Continuation of the nutrition project in Northern Newfoundland. *Journal of the American Dietetic Association*, 8, 526–31.

Various Authors. 1997. *Five hundred years of Newfoundland cookery.* St. John's, NL: Mr. Printer.

Villagarcía, M. G. (1995). *Structure and distribution of fish assemblages on the Northwest Newfoundland and Labrador Shelf.* M.Sc. thesis, Biology, Memorial University of Newfoundland, St. John's, NL.

Vitousek, P. M., Mooney, H. A., Lubchenco, J., & Melillo, J. M. (1997). Human domination of Earth's ecosystems. *Science,* 277, 494–500.

Wallace, J. (1998). Introduction. In S. Griffiths & J. Wallace (Eds.), *Consuming patterns: Food in the age of anxiety* (pp. 1–10). Manchester, UK: Mandolin.

Walters, C. J., Christensen, V. and Pauly, D. (1997). Structuring dynamic models of exploited ecosystems from trophic mass-balance assessment. *Reviews in Fish Biology and Fisheries*, 7, 139–172.

Walters, C., & Korman, J. (1999). *Salmon stocks. Background Paper No. 1999/1b.* Vancouver, BC: Pacific Fisheries Resource Conservation Council.

Wang, C., & Miko, P. S. (1997). Environmental impacts of tourism on U.S. national parks. *Journal of Travel Research,* 35, 31–36.

Walters, C., & Maguire, J-J. (1996). Lessons for stock assessment from the northern cod collapse. *Reviews in Fish Biology and Fisheries,* 6, 125–137.

Warde, A. (1997). *Consumption, food, and taste. Culinary antinomies and commodity culture.* Thousand Oaks, CA: Sage.

Watt, K .E. F., (1982). *Understanding the environment.* Boston, MA: Allyn and Bacon.

Weber, M. (1978). *Economy and society: An outline of interpretive sociology*, vol. 1, G. Roth & C. Wittich (Eds.). Berkeley, CA: University of California Press.

Weber, P. (1994). *Net loss: Fish, jobs, and the marine environment.* Worldwatch Paper 120. Washington, D.C.: Worldwatch Institute.

Welch, D. W, Ishida, Y., & Nagasawa, K. (1998). Thermal limits and ocean migrations of sockeye salmon (*Oncorhynchus nerka*): long-term consequences of global warming. *Canadian Journal of Fisheries and Aquatic Sciences*, 55, 937–948.

Wells, D. M. 1999. *Canada-Newfoundland agreement on economic renewal: A growers guide to small scale cod grow-out operations.* Prepared for Department of Fisheries and Oceans, St. John's, NL.

Whitbourne, R. (1622). *A discourse and the discovery of Newfoundland.* London: Kynston.

Wiese, F.K., Robertson, G.J., & Gaston A.J. 2004. Impacts of chronic marine oil pollution and the murre hunt in Newfoundland on thick-billed murre *Uria lomvia* populations in the eastern Canadian Arctic. *Biological Conservation,* 116, 205–216.

Wiese, F. K., & Ryan, P.C. (1999). Trends of chronic oil pollution in Southeast Newfoundland assessed through beached-bird surveys, 1984-1997. *Bird Trends,* 7, 36–40.

Wilkinson, R., & Marmot, M. (Eds.). (2003). *Social determinants of health: The solid facts.* (2nd ed.). Denmark: World Health Organization.

Williams, P. L., McIntyre, L., Dayle, J. B., & Raine, K. (2003). The "wonderfulness" of children's feeding programs. *Health Promotion International,* 18, 163–170.

Wilson, D. (2003). *Another Newfoundland scoff. A pictorial cookbook.* St. John's, NL: Flanker Press.

Wong, A. (Ed.) (2004). *First Nations nutrition and health conference Proceedings.* Conference June 2003, Squamish Nation, Vancouver, BC: Arbokem Inc.

Woodhead, A. D. (1979). Senescence in fishes. *Symposia of the Zoological Society of London*, 44, 179–205.

Wootton, R. J. (1979). Energy cost of egg production and environmental determinants of fecundity in teleost fishes. *Symposia of the Zoological Society of London*, 44, 179–205, 133–159.

World Food Summit. 1996. *Declaration on world food security.* 13–17 November. Rome: Food and Agricultural Organization (FAO) of the United Nations. http://www.fao.org/documents/show_cdr.asp?url_file=/DOCREP/003/W3613E/W3613E 00.HTM.

Worm, B, Lotze, H. K., & Myers, R. A. (2003). Predator diversity hotspots in the blue ocean. *Proceedings of the National Academy of Sciences USA,* 100, 9884–9888.

Worm, B., & Myers, R. A. (2003). Meta-analysis of cod-shrimp interactions reveals top-down control in oceanic food webs. *Ecology,* 84, 162–173.

Wright, R. (2004). *A Short History of Progress.* Toronto, ON: House of Anansi Press.

Wroblewski, J. S., Bailey, W. L., & Russell, J. (1998). Grow-out cod farming in southern Labrador. *Bulletin of the Aquaculture Association of Canada,* 98–2, 47–49.

Wu, R. S. S. (1995). The environmental impacts of marine fish culture: towards a sustainable future. *Marine Pollution Bulletin,* 31, 159–166.

Wyllie-Echeverria, S., & J. D. Ackerman. (2003). The seagrasses of the Pacific Coast of North America. In E. P. Green & F. T. Short (Eds.), *World Atlas of Seagrasses: present status and future conservation* (pp. 199–206). Prepared by the UNEP World Conservation Monitoring Centre. Berkeley, CA: University of California Press.

Yodzis, P. (1998). Local trophodynamics and the interaction of marine mammals and fisheries in the Benguela ecosystem. *Journal of Animal Ecology,* 67, 635–658.

Yodzis, P., & S. Innes. (1992). Body size and consumer–resource dynamics. *The American Naturalist*, 139, 1151–1175.

Zabel, R. W., Harvey, C. J., Katz, S. L., Good, T. P., & Levin, P. S. (2003). Ecologically sustainable yield. *American Scientist,* 91, 150–157.

GLOSSARY

Allometric rules
relationships that scale biological attributes (e.g. fecundity, mortality, metabolic rate) to organism size. Sometimes referred to as the mouse-to-elephant curve.

Anthocyanidins
flavonoid compounds found in plants such as bilberry and cranberry that have antioxidant properties. They bind free radicals in the body and have been shown to help prevent a number of long-term illnesses such as heart disease, cancer, and an eye disorder called macular degeneration (a disease of the retina that can lead to blindness). They help build strong capillaries and improve circulation to all areas of the body. They also prevent blood platelets from clumping together.

Antiscorbutic
alleviating scurvy, a disease of Vitamin C deficiency.

Aquaculture
the farming of finfish, shellfish, crustaceans and aquatic plants.

Biomarkers
signatures of groups of organisms or of certain environmental processes.

Carrying capacity
measure of habitat to sustain a population at a particular density. The abundance of individual species or groups of species will be constrained by available habitat and manipulated by past and present fishing/other human impact, natural variability and climate change.

Ecosystem services
everything not captured in market valuation such as atmospheric purification by forest and ocean plankton, the temperature buffering/climate modifying effect of the ocean, etc.

Ectotherm
"cold-blooded" animal in which the body temperature relies on sources of heat outside itself.

Extirpation	disappearance of a local population as a result of human action. The barndoor skate (*Raja laevis*) is an example of a species close to extirpation by fishing. 'Extinction' refers to the disappearance of a species from the entire planet, e.g., the dodo (*Didus ineptus*), great auk (*Pinguinus impennis*) and passenger pigeon (*Ectopistes migratorius*).
Demersal	living on or near the bottom.
Genetic introgression	process by which new genes are introduced into a wild population through backcrossing of hybrids between two populations. With aquaculture the process involves the introduction of maladaptive genes into the wild population via hybridization with farm stock.
Nauplii	larval stage of crustaceans, including sea lice. Individuals are typically very small and thus can disperse rapidly via currents during this stage.
r-K spectrum	the continuum of life history strategies from r-species with small body size, low maximum age, early maturity, high growth rate and high fecundity to K-species with large body size, old ages, late maturity, low adult mortality, slow growth and low fecundity. The term derives from the two parameters of the logistic equation, r (intrinsic population growth rate) and K (carrying capacity).
Traditional ecological knowledge (TEK)	cumulative body of knowledge, practice, and belief, evolving by adaptive processes and handed down through generations by cultural transmission, about the relationships of living beings (including humans) with one another and with their environment." (Berkes 1999: 8)
Truck system	mercantile barter system, used historically in species-scarce economies, in which goods were exchanged for other goods in the merchant store, with the merchant setting the terms of exchange. In nineteenth century Newfoundland, a quintal of fish was the equivalent of a barrel of flour.

APPENDIX

Table A.1. Scientific names of plants and animals mentioned in this book (listed alphabetically by major common or local name, within the categories: Birds; Finfish, Shellfish and Marine Invertebrates; Mammals; and Plants)

Common Names	Scientific Name
Birds	
Auk, great	*Alca impennis* (syn. *Pinguinus impennis*)
Curlew, Eskimo	*Numenius borealis*
Dovekies (bullbirds)	*Alle alle*
Duck, eider (common eider)	*Somateria mollissma*
Duck, Labrador	*Camptorhychus labradorius*
Duck, long-tailed (hounds)	*Clangula hyemalis*
Duck, northern common eider (shoreyers)	*Somateria mollissima borealis*
Eider, common (see Duck, eider)	
Eider, northern common (see Duck, northern common eider)	
Gannet, northern	*Sula bassanus*
Guillemot, black	*Cepphus grylle*
Kittiwake (black legged Kittiwake)	*Rissa tridactyla*
Murre, common	*Uria aalge*
Murre, thick-billed	*Uria lomvia*
Partridge	*Lagopus* spp.
Ptarmigan	*Lagoppus lagopus, L. mutus*
Puffin, Atlantic	*Fratercula arctica*
Razorbill (tinker)	*Alca torda*
Scoter (diver)	*Melanitta* spp.
Shearwater, greater	*Puffinus gravis*
Shearwater, sooty	*Puffinus griseus*
Snipe	*Gallinago gallinago*
Turr (Murre, common and thick-billed)	*Uria aalge* (also *Uria lomvia*)
Fish, Shellfish and Marine Invertebrates	
Alligatorfish, common	*Aspidophoroides monopterygius*
Alligatorfish, northern	*Agonus decagonus*

Table A.1. (Continued)

Common Names	Scientific Name
Fish, Shellfish and Marine Invertebrates (Continued)	
Angler, American (Goosefish)	*Lophius americanus*
Argentines	*Argentina* spp.
Black cod	*Anoplopoma fimbria*
Bocaccio	*Sebastes paucispinis*
Capelin or caplin	*Mallotus villosus*
Char, Arctic	*Salvelinus alpinus*
Clams	Bivalvia
Clam, Atlantic surf	*Spisula solidissima*
Cockle, Greenland	*Serripes groenlandicus*
Cod, Arctic or polar	*Boreogadus saida*
Cod, Atlantic or common	*Gadus morhua*
Cod, black (see Black cod)	
Cod, golden	*Gadus morhua*
Cod, Greenland	*Gadus ogac*
Cod, ling (see Lingcod)	
Cod, northern	*Gadus morhua*
Cod, rock (see Cod, Greenland)	
Crab, snow	*Chionoecetes opilio*
Ctenophore	*Mnemiopsis* spp.
Cusk	*Brosme brosme*
Dogfish, black	*Centroscyllium fabricii*
Dogfish, spiny or common	*Squalus acanthias*
Eel, American	*Anguilla rostrata*
Eel, longnose	*Synaphobranchus kaupi*
Eelpout	*Lycodes* spp.
Eelpout, Arctic	*Lycodes reticulatus*
Eelpout, Esmark's	*Lycodes esmarki*
Eelpout, Vahl's	*Lycodes vahlii*
Eulachon (ooligan, oulachen)	*Thaleichthys pacificus*
Finfish	Osteichthyes
Flatfishes	Pleuronectiformes
Flounder, winter	*Pseudopleuronectes americanus*
Flounder, witch (Sole, gray)	*Glyptocephalus cynoglossus*
Flounder, yellowtail	*Limanda ferruginea*
Flounders	Pleuronectiformes
Goosefish (see Angler, American)	
Grenadier, roughhead (also Grenadier, onion-eye or smoothspined)	*Macrourus berglax*
Grenadier, roughnose	*Trachyrhynchus murrayi*
Grenadier, roundnose, or rock	*Coryphaenoides rupestris*
Groundfish (see Cod, Flounder, Hake and other bottom fish)	

Table A.1. (Continued)

Common Names	Scientific Name
Fish, Shellfish and Marine Invertebrates (Continued)	
Haddock	*Melanogrammus aeglefinus*
Hake	*Merluccius* spp. and other genera
Hake, blue	*Antimora rostrata*
Hake, red	*Urophycis chuss*
Hake, silver	*Merluccius bilinearis*
Hake, white	*Urophycis tenuis*
Halibut, Atlantic	*Hippoglossus hippoglossus*
Halibut, Greenland	*Reinhardtius hippoglossoides*
Halibut, Pacific	*Hippoglossus stenolepis*
Herring, Atlantic	*Clupea harengus*
Herring, Pacific	*Clupea pallasii*
Humpback salmon (see Salmon, pink)	
Jellyfish, moon	*Aurelia aurita*
Ling, blue	*Molva dypterygia*
Lingcod	*Ophiodon elongatus*
Lumpfish	Cyclopteridae
Lumpfish, common	*Cyclopterus lumpus*
Lumpfish, spiny (also Lumpsucker, Atlantic spiny)	*Eumicrotremus spinosus*
Mackerel, jack	*Tracurus murphyi*
Marlin spike (also Grenadier, common; Rat-tail)	*Nezumia bairdii*
Mussel, edible blue (also wild blue)	*Mytilus edulis*
Mussel, horse also horsemussel, northern)	*Modiolus modiolus*
Ooligan or oulachen (see Eulachon)	
Perch, Pacific Ocean	*Sebastes alutus*
Pilchards	*Sardinops* spp.
Plaice, American (also Plaice, Canadian, or Dab)	*Hippoglossoides platessoides*
Poacher, Atlantic	*Leptagonus decagonus*
Pollock	*Pollachius* virens
Porbeagle	*Lamna nasus*
Redfish, deepwater	*Sebastes mentella*
Redfish, golden	*Sebastes marinus*
Redfishes, Atlantic	*Sebastes* spp.
Rock cod (see also Cod, Greenland)	*Gadus ogac*
Rockfish, copper	*Sebastes caurinus*
Rockfish, rougheye	*Sebastes aleutianus*
Sablefish	*Anoplopoma fimbria*
Salmon, Atlantic	*Salmo salar*
Salmon, Chinook (or king)	*Oncorhynchus tshawytscha*
Salmon, chum	*Oncorhynchus keta*
Salmon, coho (or silver)	*Oncorhynchus kisutch*
Salmon, humpback (see Salmon, pink)	

Table A.1. (Continued)

Common Names	Scientific Name
Fish, Shellfish and Marine Invertebrates (Continued)	
Salmon, Pacific	*Oncorhynchus* spp.
Salmon, pink (humpback)	*Oncorhynchus gorbuscha*
Salmon, sockeye or red (also Blueback, Kokanee)	*Oncorhynchus nerka*
Salmon, spring (see Salmon, Chinook)	
Sandlance	*Ammodytes* spp.
Scallop, Icelandic	*Chlamys islandica*
Sculpin, Arctic deepsea (see Sculpin, polar)	
Sculpin, polar	*Cottunculus microps*
Sculpin, shorthorn	*Myoxocephalus scorpius*
Sculpins	Cottidae
Sea lice	*Lepeophtheirus salmonis* and *Caligus clemensi*
Sea urchin, purple	*Strongylocentrotus purpuratus*
Sharks, large	Chondrichthyes
Shrimp, northern	*Pandalus borealis*
Skate, smooth	*Malacoraja senta*
Skate, spinytail	*Bathyraja spinicauda*
Skate, thorny	*Amblyraja radiata*
Skate, winter	*Leucoraja ocellata*
Skates	Rajidae
Slickhead, Baird's	*Alepocephalus bairdii*
Smelt, rainbow	*Osmerus mordax*
Snapper, red (see red snapper)	
Sole, gray (see Flounder, witch)	
Red snapper (rasphead rockfish, yelloweye rockfish)	*Sebastes ruberrimus*
Steelhead (rainbow trout, redband trout)	*Oncorhynchus mykiss*
Sucker, common	*Catostomus* sp.
Tapirfish, large scale	*Notacanthus chemnitzi*
Trout, brook	*Salvelinus fontinalis*
Trout, brown	*Salmo trutta*
Tuna, bluefin (also Atlantic bluefin, or northern bluefin)	*Thunnus thynnus*
Whelks (also Winkles, or waved whelks)	*Buccinum undatum*
Wolffish	*Anarhichas* spp.
Wolffish, Atlantic	*Anarhichas lupus*
Wolffish, northern	*Anarhichas denticulatus*
Wolffish, spotted	*Anarhichas minor*
Mammals	
Caribou (also reindeer)	*Rangifer tarandus*
Dolphins	Delphinidae

Table A.1. (Continued)

Common Names	Scientific Name
Mammals (Continued)	
Dolphin, white-sided or Atlantic white-sided (also Jumper)	*Lagenorhynchus acutus*
Fox	*Vulpes vulpes (or* syn. *V. vulpes* ssp. *fulva)*
Hare, snowshoe (also called Rabbit)	*Lepus americanus*
Killer whale or Orca	*Orcinus orca*
Lynx	*Lynx canadensis*
Marten	*Martes americana*
Moose	*Alces alces*
Orca (see Killer whale)	
Porpoise, common or harbour (also Herring jumpers)	*Phocoena phocoena*
Sea cow, Steller's	*Hydrodamalis gigas*
Sea lions	Otariidae
Seals	Phocidae
Seal, grey (Atlantic gray seal)	*Halichoerus grypus*
Seal, harbour (or harbor)	*Phoca vitulina*
Seal, harp	*Pagophilus groenlandicus*
Seal, hooded	*Cystophora cristata*
Seal, ringed (jar seal)	*Pusa hispida*
Vole, northern red-backed	*Clethrionomys rutilis*
Whale, humpback	*Megaptera novaeangliae*
Wolf, gray or common	*Canis lupus*
Plants	
Angelica, Norwegian or common	*Angelica archangelica*
Artichoke, globe	*Cynara scolymus*
Bakeapple (also Cloudberry)	*Rubus chamaemorus*
Barley (cereal grain)	*Hordeum vulgare*
Beet, beetroot	*Beta vulgaris*
Blueberries	*Vaccinium* spp.
Cabbage	*Brassica oleracea*
Carrot	*Daucus carota* ssp. *sativa*
Cedar, eastern white or northern white (also Swamp cedar)	*Thuja occidentalis*
Cedar, western red	*Thuja plicata*
Celery	*Apium graveolens*
Chick pea	*Cicer ariethinum*
Cloudberry (see Bakeapple)	
Crabapple, Pacific	*Malus fusca* (syn. *Pyrus fusca*)
Eelgrass	*Zostera marina*
Eggplant	*Solanum melongena*
Hemlock, western or Pacific	*Tsuga heterophylla*

Table A.1. (Continued)

Common Names	Scientific Name
Plants (Continued)	
Cranberry, highbush (also Squashberry)	Viburnum edule
Lettuce (garden)	Lactuca sativa
Onion (garden)	Allium cepa
Parsnip	Pastinaca sativa
Partridge berry (also Lingonberry)	Vaccinium vitis-idaea
Pepper, green	Capsicum annuum
Potato, white or Irish	Solanum tuberosum
Rape (mustard)	Brassica napus
Seaweed, red laver (edible)	Porphyra abbottiae
Soapberry (or russet buffalo-berry)	Shepherdia canadensis
Spruce	Picea spp.
Spruce, black	Picea mariana
Tuckamores (wind-stunted spruce)	Picea sp.
Turnip (garden)	Brassica rapa
Yew, Pacific	Taxus brevifolia

INDEX

D

F

Q

R